I0468984

NISTIR 5302

Evaluating Small Board and Care Homes: Sprinklered vs. Nonsprinklered Fire Protection

Scot Deal

November 1993

U.S. Department of Commerce
Ronald H. Brown, Secretary
Technology Administration
Mary L. Good, Under *Secretary for Technology*
National Institute of Standards and Technology
Arati Prabhakar, *Director*

Prepared for:
U.S. Fire Administration
Emmitsburg, MD 21727

Table of Contents

Table of Contents

List of Tables

List of Figures

VIII

Abstract

This report studied the effectiveness of sprinklered and nonsprinklered fire protection options in small Board and Care homes. The tools used to compare the effectiveness these fire protection options were mathematical fire models, experimental data and documented fire incidents. The mathematical models estimated fire protection effectiveness through a margin of safety analysis. The margin of safety is defined in this report as the excess time an evacuee has to reach a point of safety before that evacuee's exit path becomes untenable. The margin of safety calculations considered fire growth, detection/Alarm activation, evacuee egress movement and smoke tenability analysis. Two egress movement plans were simulated; one plan reflected necessary movement in a one-exit home, the second plan reflected movement in a two-exit home. Two fast-growing, large flashover fires (with high and low CO production rates) and a small, smoldering fire were modeled. Two sets of full-scale sprinklered and post-flashover fire experiments, as well as 61 documented fire incident were included in the study of fire protection system effectiveness.

The overwhelming majority of B&C fatalities occur to residents who are challenged by some disability. This disability can be mental, developmental or physical. Examination using mathematics modeling supports the use of compartmentation/evacuation in providing the same margin of safety as sprinklers during the first 15 minutes of a ground floor fire when barriers perform according to their fire-resistance ratings. Examination of the historical fire record supports sprinkler effectiveness. To date, sprinklers are more reliable and more effective than compartmentation at preventing fatalities and protecting property in the occupancy. Detection system false alarms, smoke leakage through concealed spaces, fire degradation of wall and ceiling sheathing, and resident injuries which hamper evacuation all contrive to reduce reliability of the compartmentation/evacuation system, even when doors are not chocked open.

Keywords: computer modeling, hazard analysis, board and care homes, evacuation time, detection, sprinklers, compartmentation, safety analysis., toxicity

1.0. Introduction

This study analyzed the use of sprinklered and non-sprinklered fire protection options in existing small Board and Care homes, An improvement in the level of required B&C fire safety has been suggested since the odds of dying in a fire for a B&C resident are 5 times greater than for any other residential occupancy [Life Safety Code Handbook 1991]. The reason for the higher fire risk is not that there are more fires in B&C homes, just fewer successful escapes. Most unsuccessful escapes befall residents who are mentally, emotionally or physically challenged. Two hundred eighty-eight people have died in B&C home fires in the last 23 years.

2.0. Background
2.1. History of the Current B&C Fire Safety Situation

The B&C home is a recent addition to the list of residential occupancy types. In the late 1970s and early 1980s, many individuals with mental retardation, pathologies, emotional distress or cumbersome physical challenges were released from institutions and reintroduced to the public community. This, combined with restrictions on nursing home admissions, encouraged substantial numbers of individuals with monetary constraints to live together in group housing situations. The size of these group homes varies from 3 to 1400 beds; this study looks at smaller B&C homes consisting of 16 or fewer occupied beds. Oftentimes, these small homes are converted single-family dwellings.

B&C homes are classified as occupancies apart from hospitals because B&C residents do not need constant professional care; they do however, require care above the level received by average, independent members of society. B&C homes may be thought of as "...filling the gap in the continuum between independent living and institutionalization." [Hawes 1993] The staffing level of B&C homes includes at least a night watch and typically a 24-hour watch. Nurses regularly visit the homes for check-ups and to deliver/administer medications.

During the last 22 years there have been 57 documented incidents of B&C fires resulting in one or more death. These fires represent a total of 288 deaths producing a statistical risk of dying in a fire for B&C residents 5 times higher than for individuals in other residential occupancies [LSC Handbook 1991].

2.2. Current Methods of Addressing Fire Safety in B&C Homes

There are two methods for achieving fire safety which enjoy a popular consensus in the fire protection community. These methods are compartmentation and sprinklers. The choice of which fire protection option to use is that of the owner, unless the evacuation capability of the residents is so impractical that self-evacuation becomes a doubtful prospect. In such a case, sprinklers are typically warranted as the protection method of choice. This choice is made out of recognition for the sprinkler's design capability of not only maintaining a safe egress from fire, but in controlling the fire from growing to a threatening size as well.

2.2.1. Description of Compartmentation Features

The term 'compartmentation' as used by this study implies the group of fire protection systems which maintain a separation of occupants and fire products means of physical barriers. Where it is not stated otherwise, the term compartmentation will refer to the following group of fire protection subsystems:

- fire resistant or noncombustible doors and barriers
- smoke resistant doors and barriers
- evacuation system:
    detection/alarm/notification
    evacuation paths using above construction
    points of safety using above construction
    door-enclosed stairwells using above construction
    staff and resident drill training
    communication/coordination with fire department
    evacuation evaluation

At first glance it may not be obvious why evacuation programs should be included within the compartmentation option. Without compartmentation though, there is no effective evacuation path. A compartmentation/evacuation system is needed in every building, whether sprinklered or not, because other disasters besides fire may require emergency evacuation.

Simply providing evacuation training is insufficient; training should be appropriate to the particular needs of a B&C home [Deal 1993]. Evacuation can be considered effective when people have moved to 'points of safety' in a 'timely manner.' The preferred 'point of safety' is outside the house; however, with proper construction, a point of safety may be situated inside the house. Interior points of safety should have easy accessibility for arriving residents, fire and smoke resistance for occupants [NFPA 101 §A-5-2.12.3.2, 1993], and easy access to an exit for rescue. The 'timely manner' concept is purposefully vague. Bather than define an appropriate time for every possible evacuation capability level, the 'timely manner' definition expresses the fire safety goal without placing undue restrictions on how that goal is met. For further information on B&C fire evacuation the reader is referred to references in the bibliography [Groner, 1982, Groner 1993, Levin 1992].

Successful evacuation requires performance from the detection and alarm/notification systems as well. Detection system design and installation are presented in NFPA 72. Further discussion about the proper design, installation and maintenance of these systems is beyond the scope of this project.

2.2.2. Description of Sprinkler Features

Automatic water spray sprinkler systems consist of a supply, a piping network, alarm connection and heat responsive sprinklers which allow water flow upon activation. The sprinkler system needs to be properly designed, installed and maintained to function reliably. The supply should deliver a specified flow rate of water under sufficient pressure such that if more than one sprinkler activates, enough water is present to control the fire. For a full explanation of sprinkler system performance, see NFPA 13, 13R and 13D.

2.23. Current B&C Situation

In order to design a fire safe building one must define the objective, analyze the fire hazard, then control the fire hazard. Sometimes, building fire safety codes define the fire objective, other times it is implied. If sprinklers are used to control the fire hazard, then the implied objective is that protection is provided for the room of origin and for anyone not intimately involved with the original fire. If compartmentation is used to control the fire hazard, then the implied objective is that rooms beyond the room of origin will be protected.

While there are differences among the three building codes (BOCA [Building Officials and Code Administrators 1990], UBC (Uniform Building Code 1991], SBC [Standard Building Code 1993]) and the one fire code (NFPA LSC [National Fire Protection Association 1991] [Life Safety Code 1991]), generally speaking, enclosed stairways are required for all homes that are not sprinklered. Generally speaking, sprinklers are also required in newly constructed homes and in homes that can not be evacuated in a timely manner. The BOCA and UBC tend to require more B&C fire protection features than the SBC. The most flexible and perhaps lenient code is the LSC. This code adjusts the amount of recommended fire protection based upon three different levels of B&C home evacuation capability.

The B&C home is not a tightly classified occupancy. This is partly a result of the occupancy's function as an intermediary occupancy (B&C's can range from two elderly friends living in a home to a 1,400 bed facility) and partly a result of having 50 states and one district each regulate the occupancy independently (there are 62 bureaus overseeing the licensure process in these 51 jurisdictions). These 62 bureaus recognize B&C living arrangements by more than 24 different names [Dobkin 1989]. This may seem to be an insignificant problem but an 'Assisted Living Facility' does not have to meet the same fire safety requirements that an 'Adult Residential Care' home does. There are estimated to be more than 30,000 licensed B&C homes with 500,000 beds in the United States [Hawes 1993] at the time of this publication. Conservative estimates V.S. House of Representatives 1989] place the number of unlicensed homes at another 30,000, while some reports place this number as high as 300,000 [Hill 1982]. It is not known what percentage of unlicensed homes are in noncompliance with the jurisdictional building or fire code. It is also not known what percentage of B&C residents would require evacuation assistance.

Actions available to bring unlicensed homes into regulation include fines, bans on admissions and referrals, and home closures. In a 12-month survey, 136 actions were taken against the 30,000 or more unlicensed homes by the 62 bureaus in the United States. There were at least 54 actions taken by 16 agencies to close unlicensed homes during this survey period [Hawes 1993], but this action is hesitatingly used because resources for resident relocation are scarce mall 1992, Levin 1993]. Of the 62 bureaus, 9 take an active role in locating unlicensed homes, the remaining respond to complaints and calls as they come into the office Lewin et al, 1990].

2.3. Items Influencing the Future B&C Fire Safety Situation

It is anticipated that the number of B&C type living arrangements will increase in the future. The elderly are the fastest growing segment of the American population [U.S. DoC 1989]. The need for 'defend-in-place' fire protection will increase as B&C residents age and lose their ability to evacuate in a timely manner. While the current B&C population is estimated between 1 and 1.67 million residents, it is not known how many individuals are waiting to be or want to be accepted into a B&C living arrangement. Some homes will not accept incontinent, bedfast or chair-confined individuals; other homes are simply too expensive. The high number of unlicensed facilities indicates, but is not proof of, the high demand for B&C housing.

The present trend for fire protection in residential occupancies is moving towards a national consensus-via the building and fire codes-towards adopting a sprinkler mandate [NFPA July 1993]. Several counties and provinces in the U.S. and Canada already require sprinklers in all new residential construction.

3.0 Approach-Description of Tools Used to Analyze B&C Fire Safety

The goal of this study is to compare the effectiveness of sprinkler and compartmentation fire protection options in a small B&C home. This comparison is made by using three 'measuring sticks:' mathematical fire models, full-scale fire experiments [Madrzykowski 1991, Stroup 1991], and actual B&C fire accounts [Hall 1993, Isner 1986, Isner, 1993]. The mathematical models excel at isolating influences exerted by individual fire protection features. For example, it is possible to discover whether the door assembly or the hallway smoke temperature is more influential in creating untenable bedroom conditions. On the other hand, the historical database of actual fire incidents excels at identifying which fire protection features are associated with fire fatalities. The two methods, mathematical modeling and fire incident reports, are complimentary methods for determining the effectiveness of building fire safety features at controlling fire. Experimental testing allows one to determine individual effects, just as fire modeling, albeit with a greater sense of confidence in the answers.

3.1. Computer Models as Tools for Evaluating B&C Fire Safety

Mathematical fire models were used to analyze the relative effectiveness of sprinkler performance on life safety in B&C homes because they save tremendous time and money compared to full-scale experimental testing. The fire modeling in this study is used to evaluate relative performance of individual fire safety features. Mathematical modeling is particularly well suited for relative performance evaluations because many parameters which are uncontrollable in 'real world' experiments can readily be held constant. It is true that fire models only simulate, not recreate reality, but the smoke transport model used was verified with real world-experiments [Levine 1990, Peacock, Davis et al 1991, Nelson 1992]. The sprinkler activation and suppression correlations were developed from experiment. The numerical simulations undertaken represent both severe and mild insults historically occurring in small B&C homes [Hall 1993].

The use of mathematical fire models as a tool for evaluating fire safety performance involved several submodels. The intent of the calculations from each of these submodels was to reach a point where the different building fire safety features could be analyzed from a common basis. The common basis was the margin of safety [Groner 1982]. The margin of safety is the excess time an individual has to reach a point of safety before that individual's exit path becomes untenable. The margin of safety was calculated for points along the primary and secondary egress paths inside the building. More information on how and why tenability was chosen as a basis for comparing building fire safety features is presented below under **Tenability Analysis** and **Resident Evacuation**.

The process of mathematically modeling a B&C fire scenario from established burning to untenable conditions in the exit paths and points of safety followed an outline. This outline began with choosing a design fire and home layout. Then these two items were mathematically described to the computer fire model. The computer simulated the effects of the fire by predicting the changing smoke temperatures and toxic gas concentrations at locations throughout the B&C home. The mathematical predictions are made using laws of physics and empirical correlations. The predicted smoke temperatures and gas concentrations for various rooms were then given to another mathematical model which predicted when untenability occurred. Overlaid on the timeline which predicts untenability was the time it takes a B&C resident to evacuate. The difference between the time to untenability and the time required for evacuation is the resulting margin of safety.

3.1.1. The Computer Fire Models

Different fire models were needed to simulate different phenomena. The phenomena simulated were: multi-room distribution of fire products [heat, oxygen (O_2), carbon dioxide (CO_2) and carbon monoxide (CO)], fire growth on a combustible wall surface, smoke detector activation, sprinkler fire suppression, post-flashover fire duration, untenability from smoke inhalation or heat loading and resident evacuation speed. The **CFAST1.6** [Peacock 1993] fire simulation model was used to record the growth and distribution of smoke temperatures and gas concentrations throughout the building. A wall-fire model [Quintiere 1992] was used to predict the pre-flashover growth rate of a burning wall finished with combustible wood. The **FIRE SIMULATOR** model within **FPEtool3.0** [Nelson 1990] was used to predict smoke detector activation, sprinkler activation, and in the cases of nonsprinklered fires, post-flashover fire behavior and duration. Post-sprinkler fire behavior was predicted with an algorithm developed by Madrzykowski [1992]. The N-Gas model [Levin 1985] was used to predict untenability from inhalation of smoke gases. The EGRESS model within **FPEtool3.0** was used to estimate resident evacuation speeds.

3.1.2. The Fire

The fires used in the simulations were first chosen and then mathematically described to the computer programs because **CFASI1.6** (and most other fire simulation models) is not yet capable of predicting fire growth rates. Five design fires were created and used in this study:

Table 1. Description of Design Fires

Design Fire	Fire Characteristics	Fire Location	Quantitative Fire Description
1	base case: combustible wall finish fire; sustained post-flashover burning from additional fuel sources	Ground Floor, entryway	Appendix A, B
2	same as design fire 1, but lowered the CO generation rate by a factor of 3	Ground Floor, entryway	Appendix A, B
3	same as design fire 1, but increased the CO generation rate by a factor of 10	Ground Floor, entryway	Appendix A, B
4	combustible wall finish fire, post-flashover burning but not as long a duration as design fire 1	Second Floor, hallway	Figure 2
5	smoldering fire with high CO generation rates, but never flaming combustion	Second Floor, hallway	[Bukowski 1989], pp. 2-1
6	same as design fire 1, but suppression with QR sprinklers began about 335 seconds (450 kW) after established burning	Ground Floor, entryway	Appendix B, Madrzykowski 1992

5

The base case was a very fast growing fire with a sustained flashover due to a substantial fuel load. A room capable of producing such a fire would have a fuel arrangement for rapid fire growth (drapes or Class 'C' wall finish for the 'fuse') and substantial fuel load to sustain flashover burning (book shelves, a sofa or easy chair, coffee table and reading material). The location of this first fire is shown as position 1 of Figure 1 (Figures are found after the Reference section). The second fire was identical to the first in all aspects, except for the CO production rate. In this scenario, the CO production rate was reduced by a factor of 3 to determine the impact fuel toxicity had upon the hazard posed by the fire. The third fire was the same as the second fire, except the CO production rate was increased by a factor of 10. The fourth fire was located on the second floor in the hallway and it, too, was a fast-burning flashover fire, only of short duration. The location of this design fire is shown as position 2 of Figure 1. A typical hallway capable of generating this fire would contain a fuel which burns rapidly (Class C wall finish) and one substantial fuel which contributes to the flashover duration (sofa, desk, easy chair or open closet). The significance of this fire is that it impinges upon the door. These hot conditions challenge the door smoke resistance much more than when the fire is located several rooms down the corridor or when the fire is one floor below. The fifth tire was also located in the second floor hallway and was a smoldering fire. The duration of the fires was 1/2 hour except for the smoldering fire which was 45 minutes. The 1/2 hour simulation time was chosen as a reasonable time by which the fire department could be expected to arrive and provide rescue assistance.

3.1.3. Fire Verification

To ensure the accuracy and applicability of the mathematically modeled fires to real world situations the design fires were chosen to duplicate the behavior of fires measured in full-scale experiments. The first design fire replicates a full-scale experimental burn which investigated CO formation and sought to discover how several very quick deaths occurred in a Sharon, PA [Levine 1990] townhouse fire. That fatal fire grew rapidly, fed by large quantities of wood wall finish and cabinetry in the kitchen. These fuels sustained flashover burning for a substantial length of time.

The rapid growth of the first and fourth design fire represents burning behavior of combustible wall paneling having a class C or lower rating. This rapid fire growth has been experimentally verified in full-scale room tests conducted in Sweden [Ostman 1989]. Figure 2 compares the first design fire with the Swedish room fire tests.

The mathematical models are not yet capable of accurately predicting full-scale, CO generation rates. For this reason the second and third design fires investigated the effects that CO generation rates exert upon the predicted time to untenability. The second and third design fires have the same heat release rate as the first design fire; only the CO generation rate was altered. The CO generation rates given to the mathematical model for the first design fire represented experimentally measured, full-scale, post-flashover wood fire generation rates. These generation rates are justified in Appendix A.

The fifth design fire was not experimentally measured per se, but represents a compilation of data taken on full-scale smoldering fires. Data for this smoldering fire were obtained from the first example case of the **HAZARD I** fire model manual [Bukowski 1989]. Flaming combustion was never achieved in this fire.

3.1.4. The Building

The B&C home simulated in the mathematical fire models represents a converted three-story single family dwelling. It can sleep 10 individuals in rooms located on the second and third floors. The kitchen and dining facilities are on the ground floor. The primary exit is an interior stairwell and the secondary exit is an exterior fire escape. The plan and elevation drawings for this home appear in Figure 3. The room numbers in Figure 3 depict the room number assigned by the **CFAST1.6** fire model to the data file **(see Appendix B)**. The building's walls and ceilings were sheathed with 0.013 m (1/2 in.) gypsum providing 30 minute fire resistive construction. The weather conditions simulated were summer temperatures and no wind.

3.1.4.1. Doors

The doors protecting stairways and sleeping rooms were specifically sized to conform to 'normal' doors per NFPA 80 (Fire Door and Windows). However, the smoke leakage areas around the doorways were represented differently in the CFAST1.6 model than how these leakages existed in reality. This representational change reduced computer simulation time. As seen in Figure 4, the four leakage areas around an actual door were converted into one equivalent and one exact leakage area. The one equivalent leakage area was created from the two vertical leakage areas along the hinge and latch edge and from the leakage area along the door soffit to create one equivalent vertical leakage area having the same height as the door. Heat can cause doors to bend and deflect but no deflection was modeled for NFPA 80 doors as deflections as this performance criterion is not specified. Leakage areas and dimensions of door assemblies as used in the **CFAST.1.6** model appear in **Appendix C** of this study.

Doors protecting stairwells were 0.813 m (32 in.) wide. Doors connecting bedrooms to the corridor were 0.762 m (30 in.) wide.

Smoke leakage into second and third floor sleeping rooms was simulated by leaking smoke into one room per floor. The smoke leakage into the other rooms was simulated with a vent which was equivalently sized to the summation of the remaining bedroom door leakage areas.

3.1.4.2. Stairs

The presence or absence of doors was used to vary stairwell configurations. Various stairwell configurations were examined because various stairwell configurations are allowed by the building codes. The BOCA[BOCA 1990], UBC [Uniform Building Code 1991] and SBC [Standard Building Code 1993] codes require doors at every stairwell landing in B&C homes. The LSC allows in particular circumstances stairwells without door enclosures. Typically, stairwells at the second or third floor level of 3-story homes may not be required to have doors. This open door configuration was compared to the BOCA/UBC/SBC configuration. In the mathematical fire model, the LSC configuration had doors enclosing the stairwell on the ground and second floor landings but not on the separate stairwell connecting the second and third floors.

3.1.4.3. Windows

Windows from the second and third floor sleeping rooms were modeled as open. In addition, a simulation was run where the window in the third floor bedroom was closed and the window in the third floor hallway was opened. The top of the window, the window sill and the width of the windows were 1.41 m, 0.61 m and 0.36 m (4.6 ft, 2 ft, and 1.2 ft), respectively.

Since the first design fire burned in post-flashover for a substantial duration several windows on the ground floor were modeled as breaking at the time of flashover. About the time of flashover, high levels of radiation from the fire can induce strain rates in the glass which are capable of breaking open the windows.

3.1.4.4. Sprinklers

Quick response (QR) sprinklers were simulated. The **FIRE SIMULATOR** module within FPEtool3.0 was used for predicting the sprinkler activation time; the algorithm of Madrzykowski [1992] was used to predict fire decay rates after the sprinkler activated. The QR links were modeled with a response time index (RTI) of 50 $(m/s)^{1/2}$ $[91(ft/s)^{1/2}]$ and an activation temperature of 68°C (155°F). Sprinkler system activation was simulated only with the first design fire. The sprinkler link was situated on the ceiling of the ground floor entrance room at the maximum distance allowed by the design code (NFPA 13R). This put the sprinkler performance under the most challenging conditions. A diagram of the sprinkler link placement is included as Figure 5.

3.1.4.5. Summary of Mathematical Simulation Scenarios

A summary of the building fire safety features which were mathematically modeled and analyzed for their ability to reduce/eliminate the fire hazard appear below:

A worst-case fire scenario was chosen to serve as a basis from which all further comparisons could be made. This base case scenario had a dwelling with only one exit located at the ground floor entryway, no stairwell doors, bedrooms having doors with leakage areas complying with NFPA 80 ('normal' leakage areas) and open windows to the outside. The worst-case scenario used the first design fire, which was a wall fire that rapidly grew to flashover and sustained such burning for 1/2 hour.

Table 2. Description of Mathematical Modeling Scenarios

Simulation	Fire Modeling Scenario	Design Fire	Exit Status	Figure
0	base case: no sprinklers, no stairwell doors	1	one exit; Ground Floor	15
1	stairwell doors on bottom flight only: LSC	1	two exits per Floor	16
2	stairwell doors on all stories: BOCA/UBC/SBC	1	two exits per Floor	17
3	same as Simulation 0, but lower CO generation	2	one exit; Ground Floor	18
4	same as Simulation 0, but higher CO generation	3	one exit; Ground Floor	19
5	effect of QR sprinkler	6	two exits per Floor	20
6	flashover fire in a sleeping floor hallway	4	two exits per Floor	22
7	smoldering fire in a sleeping floor hallway	5	two exits per Floor	23

Simulation # 1 was altered from the base case by enclosing the stairways with doors on the ground and second floor landing only. Doors were not installed on the separate stairwell connecting the second and third stories; this configuration is permitted by the LSC [NFPA LSC, 1991 §23-2.2.4, Exception 1]. The ground and second floor stairwell doors are labelled A and B in Figure 3. The stairwell door leakage areas conform to NFPA 80 ('normal' leakage areas).

Simulation # 2 modified the base case by installing stairwell doors at every landing.

Simulation # 3 modified the base case (Simulation 0) by decreasing the CO yield of the fire by a factor of 3.

Simulation # 4 modified the base case (Simulation 0) by increasing the CO yield of the fire by a factor of 10.

Simulation # 5 modified Simulation # 1 by observing the predicted effects of a properly designed and operating quick-response water spray sprinkler system.

Simulation # 6 modified the base case (Simulation 0) by moving the fire upstairs to the second floor hallway and reducing the duration of the flashover.

Simulation # 7 modified the base case (Simulation 0) by moving the fire upstairs to the second floor hallway and changing the fire to a smoldering type.

3.1.5. Tenability Analysis

An analysis of time to untenability was used to determine if and when a resident succumbed to the fire hazards. This point was assumed to occur when a human received either a dose of toxic gas or heat that could cause unconsciousness. The tenability limits used were the same as those used by Klote, Nelson et al [1992] and Madrzykowski [1991] and Stroup [1991].

Tenability from toxic gas exposure was modeled using the Fractional Effective Dose (FED) concept. The concept of dose assumes that if two different exposure scenarios yield equivalent toxic material concentration and exposure time products, then the doses are equivalent. Put another way, a 10 minute exposure to 5000 ppm CO would be equivalent to a 2 minute exposure to 25,000 ppm CO because the product of exposure time and gas concentration for both scenarios equals (50,000 ppm 1 min).

The FED model assumes there is a threshold dose level which a human can receive and beyond which incapacitation will occur. The FED is calculated in such a way that when the cumulative dose level exceeds the nondimensional value one-half (0.5), unconsciousness is predicted to occur. The FED model used in this study considers 3 gases: CO, CO_2 and O_2 [Levin 1985]. The model considers that the presence of CO_2 will accelerate CO uptake. Hypoxia, or O_2, deprivation, is considered without regard to the presence of CO_2. The resulting doses received from the CO/CO_2 exposure and hypoxia are added together at each timestep of the simulation to produce a cumulative dose for that timestep. The dose for the most recent timestep is added to the dose from the previous timestep to produce the cumulative dose. This cumulative dose is the FED.

TENAB, a model within **HAZARD I,** uses some specific rules regarding smoke layer height and temperature to assume whether an evacuee is willing to walk within a smoke layer or whether the evacuee will crawl to avoid the smoke. These rules were ignored for this analysis. Since all but the smoldering fire developed rapidly, the assumption is reasonable that the layer would quickly engulf residents lingering in the hall. There are limits to the FED gas model: it was developed on rat-not human-data; the FED assumes all humans respond identically to the same toxic gas doses; the FED does not account for human recovery when smoke conditions improve rather than deteriorate, the onset of lethality from gas poisoning is affected by human activity level (Figure 6); and, most importantly, the FED assumes the concept of exposure dose is valid.

Untenable conditions due to heat are predicted to occur when a gas that touches the skin has a temperature in excess of 65°C (150°F).

3.1.6. Resident Evacuation
3.1.6.1. Margin of Safety

The margin of safety [Groner 1982] is a concept used for measuring the fire hazard. Conceptually, the margin of safety eliminates the need to quantify several parameters which are highly random and difficult to predict. These parameters include time used to: recognize the alarm, investigate the fire source, retrieve personal articles, look after other evacuees' well-being, and, among others, find an escape path. The margin of safety therefore represents the amount of time, from the point of alarm activation, that an evacuee has for pursuing unpredictable and random activities. If the resident does not begin his evacuation before the end of the margin of safety, then he will be overcome by the fire before successfully reaching a point of safety (see Figure 7). The resident evacuation timeline is overlaid on the developing tenability conditions timeline to determine the margin of safety.

3.1.6.2. Detector Activation

For this study, the fire alarm is assumed to activate with the smoke detector. The smoke detector activation is predicted by an algorithm in **FIRE SIMULATOR** of **FPEtool3.0** corresponding to a rise in smoke temperature of 13°C (23°F) [Heskestad 1977].

3.1.63. Evacuation Movement

The time required for a resident to move to a point of safety depends upon the distance travelled and the movement capability. The base case scenario **(Simulation 0 in Table 2)** was used as a worst-case comparison for all subsequent simulations. As such, an exterior fire escape was assumed to NOT be available for evacuation use in the base case. For the base case it was therefore necessary that third story residents evacuate using the interior stairs, descending two flights before approaching the exterior door in the ground floor entryway **(see Figure D-1, Appendix D)**. The travel distance to this exterior door from the most remote third floor bedroom was 19 m (62 feet).

For most subsequent scenarios, the home was assumed to comply with building and fire codes. The egress time was the duration needed to move from the bed of the most remote room to the fire escape on that floor. This travel distance was 11 m or 36 feet. Travel distances are not labeled on the evacuation diagrams, but they may be obtained from cross-referencing the sample **CFAST1.6** input file **(Appendix B)** with **Figure 3** or by inspection of **Appendix D**. The room numbers in Figure 3 correspond to the room numbers in the **CFAST1.6** input file.

3.1.6.4. Travel Speeds

Travel times for 'able-bodied' residents and disabled residents also appear in **Appendix D.** The travel speeds for the disabled represent motions characterized by the Americans with Disabilities Act (ADA) [U.S. DoJ 1991]. While the ADA assumes affected individuals can not use stairs, this study assumed that travel speed on stairs would be slowed by the same proportion to which speed on level floors was slowed. The ADA assumed a travel speed 37% slower than speeds accepted for the general population. This results in foot speeds of 0.47 m/s (90 ft/min) with a 2 minute rest after each 100 feet of travel.

3.2. Historical Data as a Tool for Evaluating B&C Fire Safety

The historical database includes information from several sources: a summary report covering B&C fire incidents from years 1971 through 1992 [Hall 1993], several reports on B&C fire incidents subsequent to the publication of the summary report [Isner 1986, Isner 1993a, Isner 1993b, Isner 1993c] and two reports on full-scale fire experiments, one with post-flashover fires [Stroup 1991] and one with sprinkler-suppressed fires [Madrzykowski 1991].

The B&C fire incident reports are a study in contrasts. Hall's report contains information on fires which resulted in deaths; Isner's reports contain information on fires without deaths. Hall's summary mall

1993] gathers information on 57 fires in B&C 'homes' which resulted in the death of one or more residents. This information was obtained from NFPA's Fire Incident Data Organization (FIDO). The term B&C 'home' is noted here because most of the 57 fire incidents described in Hall's summary were not categorized, inspected and licensed as code complying B&C homes. In Hall's summary only 4% of the fire incidents involved LSC complying facilities and only 17% included facilities meeting any kind of jurisdictional safety codes at all. To obtain the 57 incidents included in the study, the FIDO system was instructed to search for and report on all fire incidents involving residential occupancies in which there existed a ' . . .pronounced mismatch between the level of care the property is designed to provide and the level of care required by some or all of the residents.'

The fire incident data from Hall's summary is divided into several categories. These categories provide insight on the effectiveness and reliability of building fire safety features scrutinized with mathematical fire modeling. While analyzing B&C fire protection features in this manner is not foolproof, it does have advantages. The computer fire models have the unique ability to hold all other building tire safety features constant while examining the effects of one particular building fire safety feature of interest. This allows one to understand the relative importance of a fire safety feature.

On the other hand, there is no mathematical fire model that calculates a reliability of individual building fire safety features on preventing fire deaths. As an example, fire modeling indicates that by opening exterior windows in upper floor bedrooms residents can accelerate the infiltration of hot, toxic smoke into the room. However effective this mechanism of smoke leakage may be, determining the reliability of the mechanism requires prediction of random human action.

While present mathematical fire models do not easily deal with issues of reliability, historical fire records do provide insight. The problem with historical records is two-fold though. First, information included in the record typically indicates only what did not work with very little information on what did work. Secondly, it can be difficult to extract the role of singular fire safety features from the nest of intradependent factors contributing towards fire development. The historical record is a valuable tool nonetheless. It provides a complimentary and balanced analysis of B&C home fire safety when used in tandem with mathematical fire modeling.

Two full-scale experimental test series were conducted in the same facility for the purpose of examining the effectiveness of compartmentation and sprinklers at mitigating the hazards of fire. Madrzykowski's test series [1991] provides a comparison of sprinklers and one-door compartmentation. Stroup's test series [1991] provides a comparison of different compartmentation designs under post-flashover conditions. The test facility layout for both test series had a bum room, a corridor, and a target room (Figure 28). The target room was outfitted with instruments which could provide a measure of smoke conditions.

In Madrzykowski's and Stroup's experiments the target room was protected by a closed door. For Madrzykowski's work this door had 'typical* leakage areas. The leakage areas of this 'typical' door were less than the leakage areas of the 'normal' door specified in NFPA 80. The leakage areas were 0.006 m (1/2 in.) on each side of the door except for the bottom, where the undercut was 0.022 m (0.87 in.). In Stroup's experiments this door assembly was interchanged with others assemblies so that in total, four door assemblies were examined. The first design retained the 'typical' door used by Madrzykowski. The second design added positive pressurization to the target room. The third design used a 'tight' door which sealed off all but the undercut of the 'typical* door. The fourth design was an accordion-style horizontally retractable door.

The fires were generated by wood cribs. Temperatures and gas concentrations were measured in all three compartments, heat release rates were recorded for the room of origin.

4.0. Results & Discussion of the Analysis of B&C fire Safety
4.1. Mathematical Fire Modeling
4.1.1. Margin of Safety

The margin of safety concept described in **Section 3.1.6.1.** was used to compare how different building fire safety features mitigate the fire hazard throughout the B&C home. Figures 8 -11 illustrate how the margin of safety varied at 4 locations within the B&C home due to different building fire safety features. The different fires used in the simulations are summarized in **Table 1**. The fires for Figures 8 - 11 were the base case fire, located on the ground floor. The vertical axis in these Figures identifies the simulation scenario. The numbers after the text descriptions correspond to the simulation number. Figures 12, 13 and 14 illustrate how changing the fire changes the smoke conditions in the B&C home.

Figures 8 - 11 have a time scale of 15 minutes while Figures 15 - 20, 22 and 23 have a time scale of 30 minutes. The 15 minute time scale was used for Figures 8 - 11 because structural response of the compartment walls is not modeled, and these surfaces were being modeled with a post-flashover heat exposure of roughly 10 minutes. After flashover, flame extension from the burning exterior of the home combined with degradation of interior wall integrity can allow smoke infiltration of upstairs rooms potentially creating untenable conditions. While 15-, 20- and 30-minute fire resistant construction is achieved in a testing facility, field construction is recognized to withstand somewhat less than this. Points in time beyond 15 minutes then (Figures 15 - 20, 22 and 23) are presented for comparing fire protection performance in idealized environments, but are not intended to be absolute measurement or predictions.

4.1.2. Tenability

Figures 15 - 20 and 22, 23 illustrate how and when untenable conditions develop for the 8 simulations described in **Table 2.** Every Figure in this series but Figure 21 has 5 subfigures (Figure 21 shows the room-of-origin exiting CO concentrations) representing conditions from the same mathematical simulation but at different locations in the B&C home. The upstairs locations are at the top of the page, the hallway locations are at the right of the page.

The different lines in Figures 15 - 20, 22 and 23 bear some explanation. There are two scales, one on the left and one on the right vertical axis. The left vertical axis quantifies the FED (Section 3.1.5.1) and the nondimensional smoke interface level, the right vertical axis quantifies smoke layer temperature.
The line representing the FED calculations always begins at the bottom left of the graph. When the FED equals or exceeds one-half, untenability is predicted. All lines depicting FED toxic-gas loading represent calculations within the upper, smoke layer. The smoke interface position appears dotted and always begins at the upper right of the Figures. The numerical value of the smoke interface height is read from the left axis. This axis has a scale which is not measured in absolute distance, but as a fraction (or percentage divided by 100) of room height. The smoke temperature is read from the right axis.

There are limitations to this mathematical study. The smoke-movement analysis assumes ideally performing, smoke-resistant barrier construction for the first 15 minutes of the post-flashover fire scenario. It also assumes that no wind or mechanical pressurization was present to affect smoke flow. Smoke entrainment in stairwells is simulated via free-standing plumes, not with inclined plumes rising along sloped ceiling surfaces that are most commonly found in stairwells.

The method by which untenable conditions develop in the modeling scenarios is predominantly through temperature (Figure 14). This is due to the fact that incapacitation is assumed to occur at 65°C (150"F), and this criterion is often passed before inhalation of incapacitating amounts of toxic gases is modeled to have occurred. This trend may not follow reality since the primary cause of death in residential fires is toxicity related. However, the 65°C criterion for incapacitation was chosen because it provides a conservative estimate of a tenability and it provides comparison with previous fire hazard studies wote, Nelson et al 1992, Stroup 1991, Madrzykowski 1991, Notarianni 1992].

4.1.3. Comparing Enclosed Stairwells and Sprinklers

In the base case simulation, which is non-complying in design because of one exit and either unenclosed stairwells or no sprinklers, the hallways rapidly became untenable (Figures 10 and 11) with the bedrooms also becoming untenable roughly 10 minutes later (Figures 8 and 9). The 10 minute delay in the onset of untenability in the bedrooms is due to closed doors. Margin of safety is different than the time to untenability. This difference is reflected by the fact that even though tenability is maintained in the bedrooms for up to 12 minutes after established burning, the margin of safety can be no greater than the tenability of the most hazardous space in the egress path, which in this case is the hallways of the respective bedrooms.

When the stairwells are enclosed (and the facility remains unsprinklered), the tenability in the hallways (Figures 10 and 11) and the bedrooms (Figures 8 and 9) is extended to 15 minutes. Evacuation of these spaces is possible at any time via the secondary exits at each floor level (Figure D1). Extrapolation of the tenability in these spaces beyond 15 minutes is not warranted because the models are not capable of simulating structural degradation of compartment walls and ceilings. This implies that the models would not account for increased smoke leakage which would occur through barrier breaches that can potentially arise from the thermal impact of the fire.

In the simulation with automatic water spray sprinklers, the simulated QR link activated at approximately 335 seconds when the fire size was about 450 kW. From this time onward, the fire size prediction followed an exponentially decay according to Madrzykowski's experimental correlation [1992]. When the small B&C home is protected by sprinklers, it is estimated that the tenability in all spaces but the room of origin will be maintained (Figure 20).

Mathematical modeling indicated that for the first 15 minutes of the fire there is little advantage to installing QR sprinklers compared with enclosing the stairways upon the life-safety of fully-mobile residents. While this may seem counter-intuitive, the results are further supported by the developing times to untenable conditions in experimentally measured, full-scale, post-flashover fire experiments.

It also may be noted that the smoke interface levels at the 2nd and 3rd stories were at the floor in the hallways and still quite low in the bedrooms. This is despite there being an open window in these bedrooms. It is not known with absolute certainty where smoke interface levels would be under such conditions, but an indication is given by measurements recorded from a full-scale, multi-storied home, post-flashover experiment wherein a bedroom with an open window was protected with a closed door]Levine, 1990]. In a series of figures depicting the smoke layer development in this bedroom, the smoke layer interface is seen to drop below the sill level before flashover occurs downstairs, and the interface level remained below 10% of the room height for the duration of post&shover burning. The large amount of smoke filling in the upstairs bedrooms is presumably caused by a chimney-like flow stimulated from the open window being located at one of the highest elevation points in the home.

4.1.4. Second Floor Hallway Flashover Wallfire

Figures 13 and 22 illustrate the effects created by a rapid growing, flashover fire on the second floor. The significance of this fire is that it allows observation of conditions inside bedrooms which are separated from the fire by only the thickness of one door. All previous Figures (8 - 11 and 15 - 20) provided observations of bedroom conditions when a fire was located one floor below and 15 m of smoke travel distance away. The doors to the bedrooms were 'normal' doors with leakage areas as permissible by NFPA 80. The fast fire development is verified in **Section 3.13** and illustrated in Figure 2.

This fire outruns the ability of the second floor smoke detector to provide adequate warning. While stairwell doors were not simulated for this scenario, additional simulations (which are not graphically illustrated) demonstrated that stairwell doors would not have prevented the rapid flashover in the hallway (Figure 22d) nor the untenable conditions in the bedrooms (Figure 22c) because sufficient oxygen was still available within the enclosed hallway for flashover. To sustain a post-flashover burn in the hallway, however, a source of oxygen would be necessary. -A bedroom with an open door and window, or an unenclosed stairway such as permitted by the L.Sc' would be sufficient to supply the necessary air.

Experimental Comparison

Stroup's full-scale experiments [1991] **(Section 4.9.1.)** showed that tenable conditions could be maintained behind a one-door compartmentation scheme for about 5 minutes when a post-flashover fire burned in a room located 13 m (43 ft) down-corridor. Toxic concentrations of smoke gases were the cause of untenability in this instance, even though 65°C was used as an incapacitating criterion. Mathematical modeling of a setup very similar to Stroup's (the fire impinged on the door rather than burned 13 m away) indicated tenable conditions could be maintained inside the room for about 5 minutes (Figure 22c) with a fire on the other side of the door. Mathematical modeling of a post-flashover fire located downstairs indicated that tenable conditions could be maintained for roughly 8 minutes in this same bedroom (Figure 1%). Much of the difference in these results stems from target room location. The greater the distance of the bedroom from the fire, the more time is required for untenable conditions to develop. In the mathematical modeling of the downstairs fire (Figure 15c), the smoke travelled approximately 15 m (50 ft) to reach the target room, In Stroup's experiments the smoke travelled 13 m (43 ft), but the target room and fire room were on the same floor (Figure 28). In addition to smoke travel distance, entrainment of air into stairwell smoke flow will extend tenability times. Entrainment dilutes toxic gas concentrations, prolonging exposures that individuals can sustain. The fact that the mathematical modeling of a one-door compartmentation scheme follows a logical and progressive order relative to the full-scale experimentally measured scheme lends credibility to the mathematical modeling.

4.1.5. Smoldering Fire

Figure 23d illustrates that the smoldering fire did not generate untenable conditions in the room of origin until 25 minutes elapsed. Untenability was eventually created by temperatures exceeding the 65°C (150°F) limit. As the smoke detector was predicted to activate at 6 minutes after ignition, the margin of safety for a resident to reach a point of safety was almost 20 minutes. From Figures 23a and 23c, it is seen that tenability is maintained in the bedrooms by simply closing the door. Tenability lasts the duration of the 45 minute simulation.

4.1.6. Fire Toxicity

Figures 14, 18 and 19 show the effect of changing the fuel toxicity. The fire is located on the ground floor for each of these simulations. Three CO generation rates are shown: base-case (this rate was

14

verified by full-scale experimental data in Appendix A), a 300% reduction and a 1,000% increase from these base-case rates. The resulting CO concentrations in the smoke emerging from the room of origin varied from 3% to 9% by volume (Figure 21). Increasing the CO concentration above the base case level of 6% increases the smoke toxicity, but not extraordinarily so because 6% CO by volume is already deadly within breaths (Figure 6). The effect of increasing the smoke toxicity from 6% to 9% decreases the time to untenability by 30R-or from 14 minutes to 10 minutes-in the third floor bedroom (Figures 15a and 19a). This effect is not displayed in Figure 14 because incapacitating conditions are predicted to occur first from the incapacitating temperature criterion of 65°C (150°F).

4.2. Historical Database

This section looks at 61 B&C home fire incidents and 13 full-scale fire experiments with the intent of determining which fire protection features are the most effective and reliable in B&C homes. The 61 B&C home fires occurred between 1971 and 1993. These fire incidents were documented by two authors and published by the NFPA. Fifty-seven of the 61 B&C home fire incidents are contained in one report [Hall, 1993]. The remaining four incidents were recorded by Isner [1985, 1993, 1993, 1993]. Nine of the 13 full-scale fire experiments examined the effects of sprinklered fire protection [Madrzykowski 1991], the other 4 experiments examined compartmentation fire protection [Stroup 1991]. All 13 experiments used the same testing facility (Figure 28).

It is not always an easy task to isolate the effects of individual fire safety features from the many factors making interdependent contributions to the growth and duration of a fire. This is especially true for fires reconstructed through historical accounts. Because of these concerns, the analysis will rely more on a global, semi-statistical approach and less upon an anecdotal review. The term 'semi-statistical' is used here because the histological method of statistically analyzing data is correct (either a building fire safety feature was present or it was not). However, it is also true that the fire incident data probably does not represent a normal sampling [Mendenhall 1992]. Data from Hall's summary includes incidents with only one or more fatality. Isner's reports include no fatalities, and all include the presence of sprinklers.

As Hall's summary contained 57 of the 61 fire incidents, his data was analyzed first. The trends which emerged from Hall's summary were double-checked with accounts from Isner and the experiments of Madrzykowski and Stroup. Trends were obtained from Hall's summary by creating a list of building fire safety features and determining in which and in how many fire incidents a particular building fire safety feature appeared. Next, the building fire safety feature was analyzed for proper implementation. Lastly a record was kept of the number of fatalities. The description of what attributes constitute appropriate implementation are found after **Table 3. Table 3** summarizes the results of Hall's analyses.

Table 3. B&C Fatal Fire Incident Summary 1971-1992 [Hall, 1993]
Fifty-seven total incidents, two hundred and eighty-eight total fatalities

	Inappropriate	Inappropriately	Appropriately	Appropriately
Sprinklers	2	4	0	0	0
Inspection, Compliance	1	4	28	214	28
Enclosed Stairwells	29[1]	87	0	46	28
Detection System	21	78	27	157	9
Licensed	14[2]	92	14	98	29
Alarm System	35	105	10	121	12
Evacuation Training	12	135	5	29	60
Combustible Finish	24	171	15	74	18

[1] this number represents homes with stairwells which did not have any doors as well as not having automatic door-closers.
[2] these facilities were not licensed.

15

4.2.1. Sprinklers

There were two incidents in Hall's survey where sprinklers were present, and these implementations were far from appropriate. The first incident had only partial sprinkler coverage. The second incident had an inadequate water supply; 4 residents died in this fire.

There are no incidents of positive sprinkler performance in Hall's summary because only incidents resulting in fatalities were included. However, Isner's reports contain several sprinkler success stories [Isner 1985, Isner 1993a, 1993b, 1993c]. Not only were there no deaths in these incidents, but substantial property was preserved as well. In an incident in Ashland, KY, the fire was controlled despite the fact that one sprinkler failed to operate due to a pipe obstruction. In an accident in Wobum, MA, a natural gas explosion in a 100-bed facility initiated what could have been a substantial property and life-loss incident (23 people were injured by the blast). Sprinklers prevented this 4-story wooden structure from being consumed by the ensuing fire and sprinklers prevented the loss of life. The fourth fire incident occurred in a hospital. It is included in this study because a fire in a sprinklered-hospital patient room is believed to behave like a fire in a sprinklered-B&C resident room. Although smoke generation necessitated the treatment of six staff members for inhalation-type injuries, no flame or heat damage occurred outside the room of origin and no patients were injured

Full-scale room experiments [Madrzykowski 1991] further verify the reduction of fire hazard by sprinklers. Nine tests monitored conditions in the bum room, adjoining corridor and a bedroom located 13 m (45 fit) down corridor. The bum room was open to the corridor; the bedroom was protected by a closed door (Figure 28). Sprinklers operating in the bum room prevented lethal conditions (as defined by this and Madrzykowski's paper) from developing in both the corridor and the bedroom, whereas nonsprinklered tests produced lethal conditions in the corridor and incapacitating conditions in the bedroom.

The two incidents where sprinklers failed to control fires in Hall's summary stress the importance of reliability. Reliability analysis tells us that it is unwise to expect proper performance from a system which works exceptionally well, but which works only part of the time. Professional inspection helps assure reliable performance.

4.2.2. Inspection, Compliance & Licensing

From Hall's data of 57 fatal fire incidents there were 2 homes with LSC' licenses and 14 homes licensed altogether. One of the LSC? homes performed as intended during the fire; the other home was actually not in compliance because of inappropriate evacuation training and penetrations in concealed-space fire walls. The twelve remaining licensed homes were inspected by state and city mental health departments. In addition, there were three homes which were not licensed that nevertheless accepted patients through administrative offices of state hospitals and mental facilities. These historical data showed little advantage for obtaining an operating license. About equal numbers of residents died in licensed as in unlicensed homes.

For half of the homes in Hall's summary there were no inspection records. Nevertheless, threequarters of all 288 deaths occurred in the uninspected half of the homes. For those B&C homes which are inspected, often times the inspectors are nurses, health examiners, welfare workers or other agents of the state health or social security offices [Hawes 1993]. Significant room for improvement exists in the area of inspecting B&C homes for fire safety and how subsequent compliance can be enforced.

4.2.3. Enclosed Stairwells

'Appropriate implementation' of this feature required automatic door-closers to be in place at all stairwell entrances. There was not one case among the 57 incidents which had reference to full compliance with this feature. There were four homes among the 29 incidents with 'inappropriate implementation' which had stairwell door closers at one, but not the other level of a flight of stairs. At other stairwell flights there was either no doors or no door-closers. Furthermore, one of these four 'inappropriately implemented' homes was responsible for 31 of the 46 deaths. This particular incident had other contributing features to account for the high death toll in addition to the nonconforming set of stairwell enclosures. The other contributing features were combustiblewall finish and a locked door.

The column labeled 'Incidents without Reference to the Feature' included eleven fire incidents in one-story homes.

4.2.4. Detection Systems

The term 'inappropriate implementation' for detection systems meant either the smoke detector itself or its power source was missing. Lack of a power source was most often associated with battery operated detectors.

Appropriate detector system performance was not critically associated with improvement to life safety as there were slightly more incidents and deaths associated with 'appropriately implemented' detection systems as there were 'inappropriately implemented systems. The poor correlation of detection performance is presumably associated with building evacuation capability. While evacuation capability is explained in detail below, the essence (Sections 4.6 and 4.7) is that B&C residents who are dying in f&z, are much more likely to be inappropriately trained in evacuation procedures and disadvantaged with regard to evacuation movement.

Existing detection systems can be improved by powering them from the house electrical system, adding detectors to the attic space for early warning of concealed space fires, and linking detector stations together. Proper detector location can extend the margin of safety. Concealed space fires are dangerous because they can grow quite large before being detected. Detectors are not required in B&C home attic spaces but 7 incidents involved fatal fires in attics or other concealed spaces. These 7 fires accounted for 68 of the 288 fire related deaths. A significant number of fires (15) also originated in common areas such as living rooms or lounges on the ground floor. Smoking was the primary -ignition source. Detectors located in or adjacent to these common areas would identify these and other high hazard areas. Also, testing the operating status of each detector on a rotating basis by initiating evacuation drills with different detectors would provide an additional method of insuring proper system maintenance.

4.2.5. Alarm Systems

An 'appropriate implementation* for an alarm system simply meant that an alarm system was installed and functioning. Proper alarm system performance, like detection system performance, was not highly associated with life safety, presumably for the same reasons mentioned above: poor evacuation training or an inability of the residents to recognize the signal. Existing alarm systems may be improved by automatically transmitting an activation signal to the fire department. Using individual alarm stations as a means of starting evacuation drills can be an additional method of verifying system operation.

4.2.6. Evacuation Training

'Inappropriate implementation' of evacuation training occurred in 12 incidents. Here, the training either reinforced bad habits or simply ignored poor performance exhibited by the residents. The significance of appropriate evacuation training is even greater than what appears at first glance. This is because a single incident was responsible for 25 of the 29 deaths associated with B&C homes having 'appropriate implementation' of evacuation training. The only reason for deeming the evacuation training of this singular incident as 'appropriate' was the common-sex&al instructions that discouraged residents from using a window access to the secondary, exterior fire escape. Most residents in this home were elderly and physically disabled. What was inappropriate for this incident was insufficient access to egress. Most of the 25 residents died upstairs as smoke prevented their progress down the primary interior stairwell.

From the trends in **Table 3** it is evident that 'appropriately implemented' evacuation programs significantly improve B&C life safety. Conversely, inappropriate evacuation programs strongly correlate with fire deaths. It may be unreliable to assume that B&C evacuation programs will be appropriately maintained without professional inspection and supervision. The question as to whether challenged B&C residents would benefit from evacuation training programs is uncertain. Even though drill evacuation times were observed to be 'short' and independent of resident 'adaptive behavior,' extrapolating this performance to actual emergencies was a conclusion the researchers would not make Barney-Smith 1978].

The two most traumatic and frequently cited factors in inappropriate evacuation programs were: failing to install compensating fire protection features in response to deteriorating evacuation performance, and failure to tram residents in the use of alternative exit paths when their preferred exit path was blocked. When evacuation performance deteriorates the home needs to compensate with additional fire protection. This can be accomplished with either sprinklers or a second, protected exit path. Residents can be trained to seek an alternative exit route out of the building with proper reinforcement. Factors to consider when evaluating a home evacuation program include: testing individual detection and alarm stations for operability and signal effectiveness (this can be done in conjunction with evacuation drills), determining if residents waste time by gathering personal belongings, dress, or attempts to fight the fire with hand-held extinguishers. Lastly, it would be very instructive for both the B&C home and the local fire station to cooperate in planning and practicing evacuation drills. Additional evacuation advice for B&C homes is cited in the bibliography [Groner 1992].

4.2.7. Evacuation Capability

In reading through the above fire safety features many may have appeared superficial in the sense that a healthy adult in a typical home would escape a fire regardless of whether there were properly installed fire safety features or not. Ordinary residents of typical homes will successfully execute an unrehearsed exit before perishing in a fire. 'This is exactly the point; residents of B&C homes are not ordinary or typical. B&C residents who are dying in fires are disadvantaged.

The B&C resident that is typically dying in fires may be mentally or physically disadvantaged, or simply slow in movement from old age. Of the 57 fire incidents in Hall's summary, 54 incidents involved residents who were in some manner, challenged.

There were three incidents unaccounted for in **Table 4** which, although anecdotal, remain instructive nonetheless. These three incidents involved residents having significantly higher evacuation capabilities than typical B&C residents; these residents were released prisoners living in half-way houses. Except for a lone victim who suspiciously fell to his death from a 5th floor, the actions of the survivors support the case that handicapped-not ordinary-residents are dying in fires. In the first incident, three residents of a half-way home escaped from second floor windows. In the second incident, one survivor climbed out of her window while another rappelled from the third floor using bedsheets she tied together. This study is not advocating bedsheet tieing as a practice for fire evacuation, rather it is stating that the B&C residents who have historically died in fires do not have the same evacuation capabilities as the general population. B&C residents with poor evacuation performance require enhanced fire protection systems to survive fires.

Table 4. Summary of Resident Capability in 57 Fatal B&C Fires [Hall 1993]

Resident Capability	Number of Fatal Fire Incidents
Mentally Retarded	24
Former Mental Ward Patients	8
Physically Disabled	4
Elderly (+ 60 years)	18
No Information Available	3
Total Number of Fatal Fires	57

Staffing is another form of capability that is important to B&C fire evacuation. In 4 of the 57 fire incidents there was no staff present to assist evacuation. A recent study found that staffing levels infrequently '..addressed the individual needs of residents,' particularly wheelchair or bedfast residents [Hawes 1993].

4.2.8. Combustible Wall/Ceiling Finish

A very high correlation between combustible finish material and life safety exists in **Table 3**. Thirteen of the fifteen largest life-loss fires involved combustible finish material; more than 170 of the 288 deaths occurred in incidents having some form of combustible wall or ceiling finish material.

Wall finish is clearly one of the major contributors to large life-loss fires in B&C homes. This building feature has more life-loss associated with it than any other category. 'Appropriate implementation* of this building feature might include an accompanying sprinkler system installation.

4.2.9. Sprinklers versus Compartmentation

The interdependencies among building fire safety features are evident from the preceding sections. The focus now turns to the simpler and primary task of comparing sprinklers to compartmentation. The reader is reminded to refer to **Section 2.23** for a definition of compartmentation and sprinklers as they will be used in the following discussion.

4.2.9.1 Effectiveness

The effectiveness criterion considers how well a fire protection system performs when it operates perfectly. The reliability feature considers how likely it is that a system will act perfectly. There are

---

1. Large life-loss fire is used here to imply an incident with 4 or more fatalities.

three sources of information (modeling, experimentation, fire incident histories) and a logical argument which justify that sprinklers provide more effective fire protection than compartmentation.

A logical argument says that an idealized sprinkler system provides higher life safety and property protection than an idealized compartmentation system because sprinklers address the fire hazard more fundamentally: sprinklers limit the fire, compartmentation limits the fire's effect. Sprinklers protect, to a degree, the room of origin and those people in the room not intimately involved in the fire; compartmentation does not protect the room of origin. There is some concern about the damage caused by water sprinklers, but fire damage caused by a lack of sprinkler protection should also be of concern.

The mathematical modeling demonstrated that an idealized enclosed stairwell compartmentation scheme can preserve life safety in upstairs floors for periods up to 15 minutes; QR sprinklers can protect upstairs rooms as long as water is available. Mathematical modeling also suggested that a typical single closed-door compartmentation scheme would protect a bedroom for 4 minutes when a post-flashover fire burned on the opposite side of the door.

Full-scale experimental measurements demonstrated that sprinklers provide superior protection to compartment&ion in reducing the hazard in corridors and the room of fire origin [Madrzykowski 1991]. These same experiments, however, showed that a one-door compartmentation scheme provided post-flashover protection to the 'downcorridor target room' for periods up to 6 minutes [Stroup 1991] and 20 minutes under sub-flashover fire exposure [Madrzykowski 1991]. In these experiments, the fire was not burning on the opposite side of the target room door, but rather in a room located 13 m (43 ft) away down a connecting corridor.

The historical data listed 6 incidents where sprinklers were installed and maintained as designed There were no deaths in these incidents and the property damage was less severe than would have been the case had compartmentation been the protection system.

4.2.9.2 Installation, Reliability and Life Expectancy

The advantage of compartmentation is that many components of the system are already in place within existing homes. Sprinkler installation for existing homes requires retrofit work: a relatively time-consuming approach compared with new construction installation. Both sprinklers and compartmentation are designed to serve for the duration of building use.

The compartmentation system involves barriers, doors, detection/alarm, and evacuation training. Each of these sub-systems must be maintained in order to keep the entire system reliable. Numerous holes and duct penetrations accumulate in walls as a result of home improvement and renovation projects. These, together with the unavoidable presence of small cracks and clearance tolerances which are even present in new construction, accumulate and can become significant paths for smoke leakage in existing structures.

While compartmentation can provide safety that is comparable with sprinklers for short durations after flashover in small B&C homes, the question really is, what is the likelihood that the compartmentation system will be maintained? It is an easy exercise to patch holes in walls and ceilings, remove door locks and door chocks, and to replace missing/discharged smoke detector batteries, but this maintenance requires human effort. Even if the compartmentation features were maintained it is a safe bet that it will be even harder to maintain resident evacuation capabilities. The B&C population is growing, and growing old in place. It is possible for an evacuation program which was appropriate 5 years ago to no longer be appropriate today. Sprinklers on the other hand circumvent the necessity for fire evacuation; and maintaining a non-moving, non-human sprinkler system is easier than maintaining a movable, human, multi-component system.

Sprinkler system reliability is relatively easy to maintain once the system passes its performance test. This is because the system is rarely used and hence rarely disturbed from working order. The reliability of sprinklers is not perfect though. Insurers place the field reliability of sprinklers in the 92-97% range binder 1993]. The historical B&C fire data reviewed in this study agrees with these figures. Only one fatal fire incident among 57 during the last 22 years resulted from a sprinkler system failure, (the other 56 incidents resulted from compartmentation/evacuation system failures). Most sprinkler system failures are attributable to a water supply problem. To avoid these problems, owners should verify proper water discharge from testing stations.

5.0. Conclusions

The conclusions from this study are categorized into those obtained from mathematical modeling and those obtained from the historical database. The conclusions obtained from mathematical modeling are verifiable to the extent that four full-scale, post-flashover experiments of closeddoor smoke leakage support trends indicated by the tenability analysis. The conclusions are furthermore verified in that the mathematical smoke movement predictions were checked for accuracy against smoke movement measurements from multi-compartment, full-scale experiments. The automatic sprinkler detection and sprinkler fire suppression predictions were also verified with full-scale experimental measurements. The design fires, too, were based upon experimentally measured and documented sources. Conclusions obtained from the historical database are verifiable to the extent that third-hand accounts of fire incidents occurring over the past 21 years can be used to accurately extract the influence individual fire safety features exerted upon fatal fires incidents. Regardless from which source the results were obtained, they apply almost exclusively to detached, structures up to three-stories in construction because this type of B&C building was the singularly dominant structure contained in the database.

From the mathematical modeling:

- Automatic sprinklers when properly designed, installed and maintained provide superior fire protection not only because they maintain tenable conditions for safe evacuation, but even more importantly, because they address the hazard at its source; sprinklers suppress the fire.

- Compartmentation or an automatic sprinkler system control the hazards of a downstairs fire from reaching fully-mobile residents for a period up to 15 minutes. Compartmentation, as it is used here, refers to all systems that maintain a properly designed and installed barrier between the residents and fire products. These systems include wall and ceilings, doors, enclosed stairwells, detection/alarm systems and evacuation training.

- Smoldering fires did not pose untenable conditions due 'to temperature or toxicity in the room of origin for 25 minutes, almost 20 minutes after smoke detector activation. The life safety threat from the smoldering fire was extremely small for residents capable of placing a closed door between themselves and the fire.

From the historical fire incident database:

- Most B&C fire deaths have occurred to mentally, physically or developmentally disabled residents.

- A very high correlation of fire fatalities is associated with B&C homes containing combustible finish material and lacking automatic fire sprinklers.

- Significant reductions in fire deaths were associated with B&C homes which maintained appropriate evacuation programs.

- Requiring improved fire protection systems may not improve B&C fire safety without better fire inspection, licensure and enforcement methods.

6.0. Summary

Mathematical modeling and experimentation tell us that automatic sprinklers are the most effective solution for protecting life and property in small, existing Board and Care homes. Typical Board and Care compartmentation, when appropriately constructed and maintained, will provide life safety for modest durations after flashover. Compartmentation effectiveness depends on the ability of residents to react and escape in a timely manner, a process whose absence has led to most of the observed Board and Care fatalities. The historical fire record tells us that the Board and Care residents who are typically dying in fires are challenged by mental, developmental or physical disabilities. A large number of the deaths in B&C homes are associated with fires involving combustible wall finish. B&C homes maintaining appropriate evacuation programs had noticeably reduced fire fatalities. Current estimates on the portion of unlicensed Board and Care homes range from 50 to 75 percent.

7.0. Acknowledgements

Thanks are extended to Mr. Maruskin and Mr. Coyle and the U.S. Fire Administration for supporting this project.

8.0. References

Babrauskas, V., Harris, R.H., et al., "The Role of Bench-Scale Tt..i Data in Assessing Real-Scale Fire Toxicity," NIST Technical Note 1284, Gaithersburg, MD 20899, January 1991.

Building Officials and Code Administrators International, Inc., 1990 Edition, Country Club Hills, IL 00478-5795, Table 916.1, pp. 194.

Bukowski, R-W., Peacock, R.D., "Example Cases for the HAZARD I Fire Hazard Assessment Method," NIST Handbook 146, Volume JII, NIST, Gaithersburg, MD, June 1989. .

Deal, S., "Small Board and Care Fire Evacuations: A Guide for the Fire Safety Professional," NISTIR 5222, MST, Gaithersburg, MD 20899, July 1993.

Dinenno, P.J. editor-in-chief, et al. "SFPE Handbook of Fire Protection Engineering, 1st. Edition," National Fire Protection Association, Quincy, MA 02269, 1988.

Dobkin, L. "The Board and Care System: A Regulatory Jungle," The American Association of Retired Persons, Washington, D.C., 1989.

Groner, N.E., "Matter of Tie-A Comprehensive Guide to Fire Emergency Planning," NBS-GCR-82-408, MST, Gaithersburg MD 20899, November 1982.

Groner, N.E., "A Guide to Board and Care Fire Safety Requirements in the 1991 Edition of the Life Safety Code," NIST-GCR-93-629, MST, Gaithersburg, MD 20899, July 1993.

Hall, J.R., "U.S. Fires in 'Board and Care' Homes Matrix Display of Selected Fatal Fires. Special Analysis," NIST-GCR-93-627, MST, Gaithersburg, MD, April 1993.

Hawes, C., Wildfire, J.B., Lux, L.J., "The Regulation of Board and Care Homes: Results of a Survey in the 50 States and the District of Columbia," American Association of Retired Persons-Public Policy Institute, Washington, D-C., 1993

Heskestad, G., Delichatsios, M. "Environments of Fire Detectors-Phase 1; Effects of Fire Size, Ceiling Height and Material, Volume II-Analysis," Technical Report Series No. 22427, RC 77-T-11, Factory Mutual Research Corporation, Norwood, MA, 1977.

Hill S., Best, R., "Fires in Two Boarding Facilities Kill 34 Residents," Fire Journal, National Fire Protection Association, Quincy, MA, 76(4): 44-49, July 1982.

Isner, M., Smith, R., "Fire in Boarding Home: A Success Story,' Fire Journal, Vol 80, No 2., March 1986.

Isner, M. "Nursing Home Fire, Sprinkler Success, Ashland, KY, June 2, 1993," National Fire Protection Association, Quincy, MA 02269, 1993.

Isner, M., "Nursing Home Fire, Woburn, MA, October 30, 1992," National Fire Protection Association, Quincy, MA 02269, 1993.

Isner, M., "Hospital Fire, Sprinkler Success, Weymouth, MA January 24, 1993," National Fire Protection Association, Quincy, MA 02269, 1993.

Klote, J-H., Milke, J.A. "Design of Smoke Management Systems," ASHRAE, Atlanta, GA 30329, 1992, pp. 22.

Knox, F.S., Francis S, Knapp, S.C. "Human Response to Fire," AGAARD Lecture Series No. 123, Neurilly Sur Seine, France, 1981.

Levin, B.C., et al., "Toxicological Effects of the Interactions of Fire Gases and Their Use in a Toxic Hazard Assessment Computer Model," The Toxicologist, 5, 127, 1985.

Levin, B.M., Groner, N-E., Paulsen, R., "Affordable Fire Safety in Board and Care Homes. A Regulatory Challenge-Final Report, " NIST-GCR-93-632, NIST, Gaithersburg, MD 20899, July 1993.

Levin, B.M., Groner, N.E., "Human Behavior Aspects of Staging Areas for Fire Safety in GSA Buildings," NIST-GCR-92a, NIST, Gaithersburg, MD 20899, April 1992.

Levine, R.S., Nelson, H-E., "Full Scale Simulation of A Fatal Fire and Comparison of Results with Two Multi-room Models," NISTIR 90-4268, NIST, Gaithersburg, MD 20899, August 1990.

Lewin/ICF Inc., James Bell Associates, "Descriptions of and Supplemental Information on Board and Care Homes Included in the Update of the National Health Provider Inventory," U.S. Department of Health and Human Services, Office of the Assistant Secretary for Planning and Evaluation, Washington, D-C., 1990.

Life Safety Code Handbook, "National Fire Protection Association, Section 21-1.1 pp. 684, 1991.

Linder, K.W., "Field Reliability of Fire Suppression Systems," invited lecture at the Workshop on Balanced Design Concepts, NIST, Gaithersburg, MD, June 30 - July 2, 1993.

Madrzykowski, D., "The Reduction in Fire Hazard in Corridors and Areas Adjoining Corridor Provided by Sprinklers," NISTIR 4631, NIST, Gaithersburg, MD 20899, July 1991.

Madrzykowski, D., Vettori, R.L., "A,Sprinkler Fire Suppression Algorithm for the GSA Engineering Fire Assessment System, " NISTIR 4833, MST, Gaithersburg, MD 20899, May 1992.

Mendenhall, W., "Statistics for Engineering and the Sciences, 3rd Ed.," MacMillan Publishing, New York, NY 10022, 1992, pp. 174.

Moritz, A.R., Henriques, F.C., "Studies of Thermal Injury. Part 2. The Relative Importance of Tie and Surface Temperature in the Causation of Cutaneous Burns," American Journal of Pathology, vol. 23, (5), 1947, pp 695-720.

National Fire Protection Association Life Safety Code', 1991, Section 22-2.2.4, Exception 1, Quincy, MA 02669, February 8, 1991.

National Fire Protection Association Life Safety Code', Quincy, MA 02669, 1993 code pending acceptance.

National Fire Codes Newsletter, National Fire Protection Association, Quincy, MA 02269, July 1993.

Nelson, H.E. "FPEtool: Fire Protection Engineering Tools for Hazard Estimation," NISTIR 4380, MST, Gaithersburg, MD 20899, October 1990.

Nelson, H.E., Deal, S. "Comparing Compartment Fires with Compartment Fire Models," Proceedings of the 3rd International Symposium on Fire Safety Science, Edinburg, Scotland, 1992.

Notarriani, K.A., "Report of Test FR 5982," MST, Gaithersburg, MD 20899, October, 1990.

Ostman, B., Nussbaum, R., "Correlation Between Small-Scale Rate of heat Release and Full-Scale Room Flashover for Surface Linings," 2nd International Symposium of Fire Safety Science, Hemisphere Publishing, New York, 1989, pp. 823-832.

Peacock, R.D., Davis, S., Babrauskas, V., "Data for Room Fire Model Comparisons" Journal of the National Institute of Standards and Technology, Vol 96, (4) July 1991.

Peacock, R.D., Forney, G.P., Reneke, P., Portier, R., Jones, W.W., " CFAST, the Consolidated Model of Fire Growth and Smoke Transport," MST Technical Note 1299, MST, Gaithersburg, MD 20899, February, 1993.

Quintiere, J.G., Cleary, T., "A Simulation Model for Fire Growth on Materials Subject to a Room-Comer Test," Fire Retardant Chemicals Association, Orlando, FL, March 29-April 1, 1992
Standard Building Code, Southern Building Code Congress International, Inc., Birmingham, AL, 35213-1206,1993

Ramey-Smith, A.M., Fechter, J.V., "Group Homes for the Developmentally Disabled: Case Histories of Demographics, Household Activities, and Room Use," NBSIR 79-1727, MST (formerly NBS), Gaithersubrg, MD 20899, 1979.

Stroup, D.W., Madrzykowski, D., "Conditions in Corridor and Adjoining Areas Exposed to Post-Flashover Room Fires," NISTIR 4678, MST, Gaithersburg, MD 20899, September 1991.

Uniform Building Code, International Conference of Building Code Officials, Whittier, CA 90601, 1991.

U.S. Department of Commerce, Bureau of the Census, "Census Catalog and Guide, United States," U.S. Government Printing Office, 1989.

U.S. Department of Justice, "Nondiscrimination on the Basis of Disabiity by Public Accommodations and in Commercial Facilities: Americans with Disabilities Act 28 CFR Part 36, Final Rule, Federal Register, July 26, 1991.

U.S. House of Representatives, Chairman, Subcommittee on Health and Long-Term Care of the Select Committee on Aging; 101st Congress, First Session, "Board and Care Homes in America: A National Tragedy," U.S. Government Printing Office, Committee Publication No. 101-711, March 1989.

9.0 Bibliography

Altman, I., Taylor, D.A., Wheeler, L., "Ecological Aspects of Group Behavior in Isolation," Journal of Applied Social Psychology," 1971, 1, pp. 76-100.

American Society of Testing and Materials, Building Seals and Sealants; Fire Standards; Building Construction, Volume 04.07, Philadelphia, PA 19103-1187, 1990, pp 321-7.

American Society of Testing and Materials, Standard on Building Economics; E06.81, Philadelphia, PA 19103-1187, 1992.

Babrauskas, V., Levin, B.C., Gann, R.G., "A New Approach to Fire Toxicity Data for Hazard Evaluation," ASTM Standard News, 14, 28-33, 1986.

Bennet, C.A., "Man-Environment Interactions: Evaluations and Applications," Stroudsburg, PA, Dowden, Hutchinson, & Ross, 1975.

Berensen, P.J., Robertson, W.G., Bioastronautics Data Book, 2nd Ed.. NASA SP-3006, Chapter 3, NASA, Washington, D.C., 1973.

Blockley, W.V., in "Biology Data Book," Federation of American Societies for Experimental Biology, NIH, Bethesda, MD, 1973.

Brill, M., "Using Office Design to Increase Productivity, Vol. I&II, Workplace Design and Productivity, Buffalo, NY, 14214, 1984.

Buettner, K., "Effects of Extreme Heat on Man," Journal of American Medical Association, 28 October, 1950, pp 732.

Bukowski, R.W., "Analysis of the Happyland Social Club Fire with HAZARD I," Arson & Investigator, 42, 3, 3647, March 1992.

Cascio, W.F., "Costing Human Resources: The Financial Impact of Behavior in Organizations," New York: Van Nostrand Reinhold, 1982.

Chapman, R-E., "Cost Estimation and Cost Variability in Residential Rehabilitation, NBS Building Science Series 129, MST, Gaithersburg, MD 20899, November 1980.

Groner, N.E., Levin, B.M. "Human Factors Considerations in the Potential for Using Elevators in Building Emergency Evacuation Plans," NIST-GCR-92-615, MST, Gaithersburg, MD 20899, September 1992.

Klote, J.H., Nelson, H.E., Deal, S., Levin, B.M., "Staging Areas for Persons with Mobility Limitations," NISTIR 4770, MST, Gaithersburg, MD 20899, February 1992.

Knox, F.S., Knapp, S.C., Wachtel, T.L., "Mathematical Models of Skin Burns Induced by Simulated Postcrash Fires as Aids in Thermal Protective Clothing Design and Selection," U.S. Army and Aeromedical Research Laboratory Report No. 78-15, Fort Rucker, Alabama, June 1978.

Lippiatt, B., Weber, S., "Productivity Impacts in Building Life-Cycle Cost Analysis," NISTIR 4762, MST,

Gaithersburg, MD 20899, February, 1992, pp 1-3.

National Fire Protection Association 252, Standard Methods of Fire Tests of Door Assemblies, 1990 Edition, Section 6-2.5 and 6-2.1.

Peacock, R-D., Jones, W .W., Bukowski, R.W., Forney, C.L., "Technical Reference Guide for the HAZARD I Fire Hazard Assessment Method, Version 1.1," MST Handbook 146, Vol. II, MST, Gaithersburg, MD 20899, June 1991.

Perry, R.H et al., Chemical Engineer's Handbook, 5th Ed., McGraw-Hill Book Co., New York, NY, 1973.

Purser, D.A. "Toxicity Assessment of Combustion Products," SFPE Handbook of Fire Protection Engineering, NPPA, Quincy, MA 02269, 1992.

Rawie, C.C., "Estimating Benefits and Costs of Building Regulations, A Step by Step Guide," NBSIR 81-2223, MST, Gaithersburg, MD 20899, June 1981.

Ruegg, R.T., Fuller, S.K., "A Benefit-Cost Model of Residential Fire Sprinkler Systems, NBS Technical Note 1203, MST, Gaithersburg, MD, 20899, 1984.

Ruegg, R-T., "Least-Cost Energy Decisions for Buildings, Introduction to Life-Cyle Costing," MSTIR 4309, MST, Gaithersburg, MD 20899, April 1990.

Schmidt, E., Lt., written memorandum on sprinkler costs, Prince Georges Fire Department, CAB, Upper Marlboro, MD, August 1993.

Simms, D.L., Hinkley, P.L., "Fire Research Special Report No. 3 on Heat Loading and Untenability," Her Majesty's Stationary office, London, 1963.

Figures

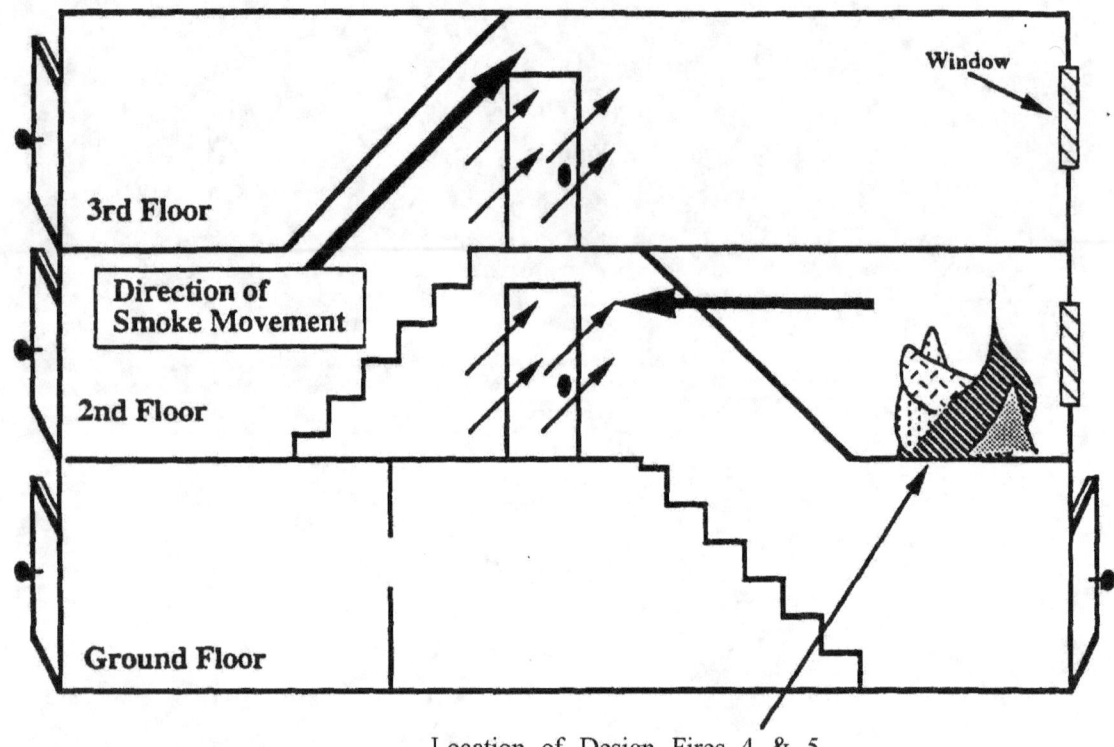

Location of Design Fires 4 & 5

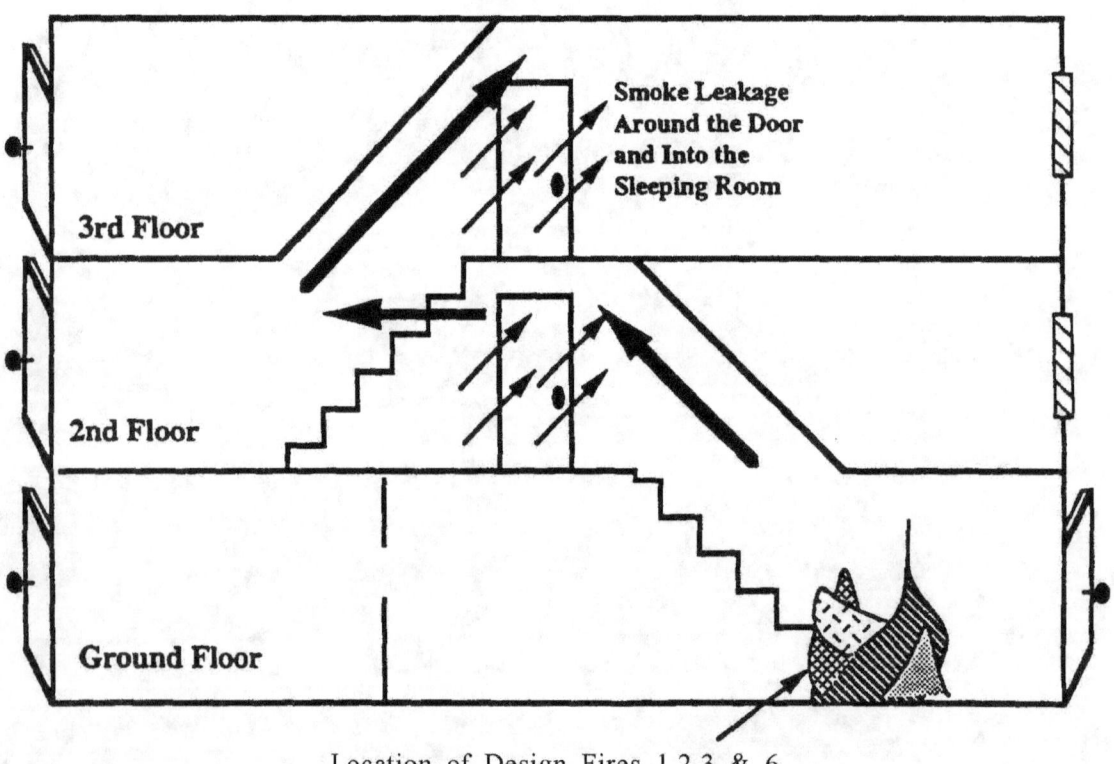

Location of Design Fires 1,2,3 & 6

Figure 1. Elevation View of Fire Locations

30

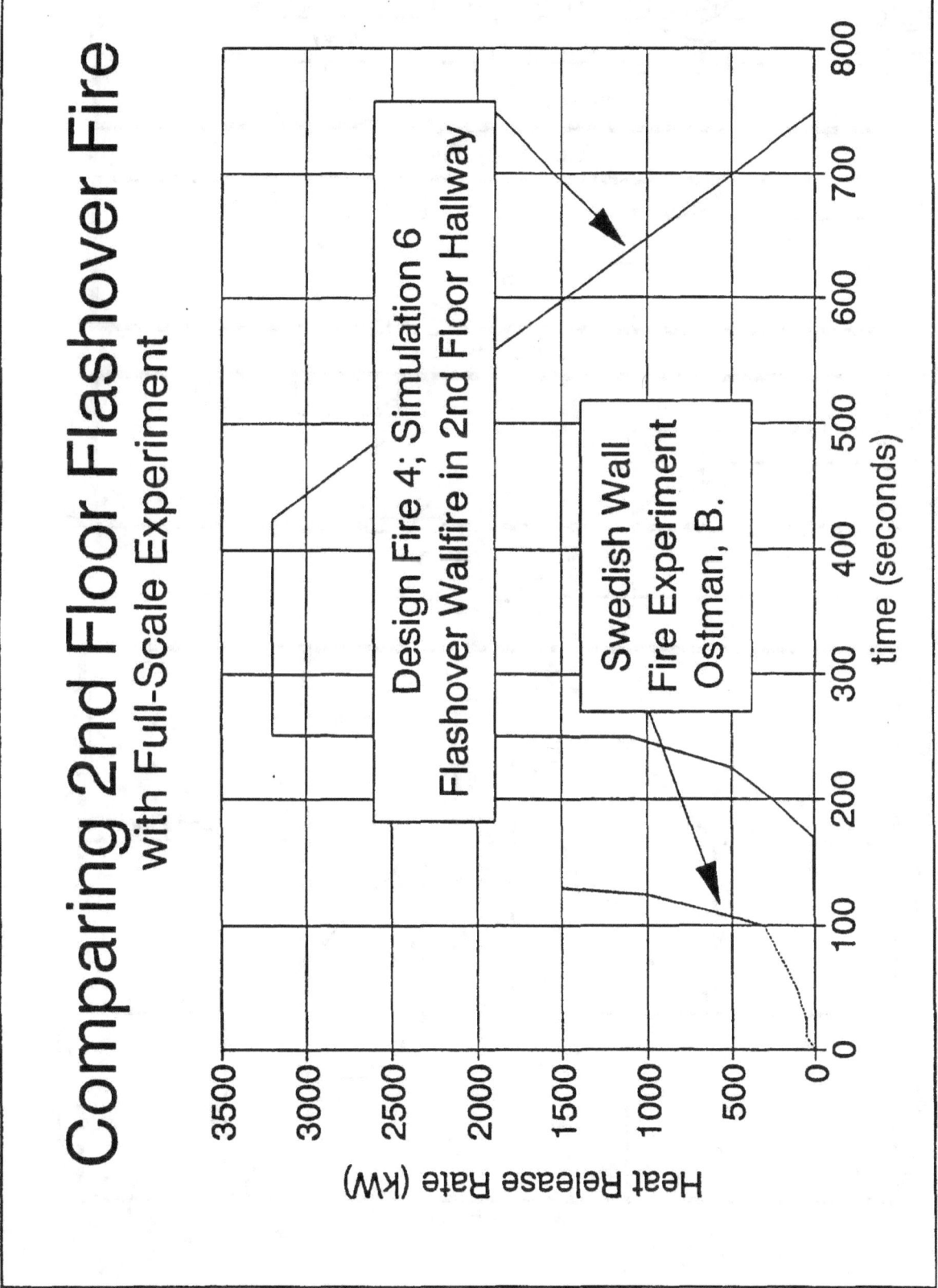

Figure 2. Verifying the 2nd Floor Flashover Fire with Full-Scale Experiment

31

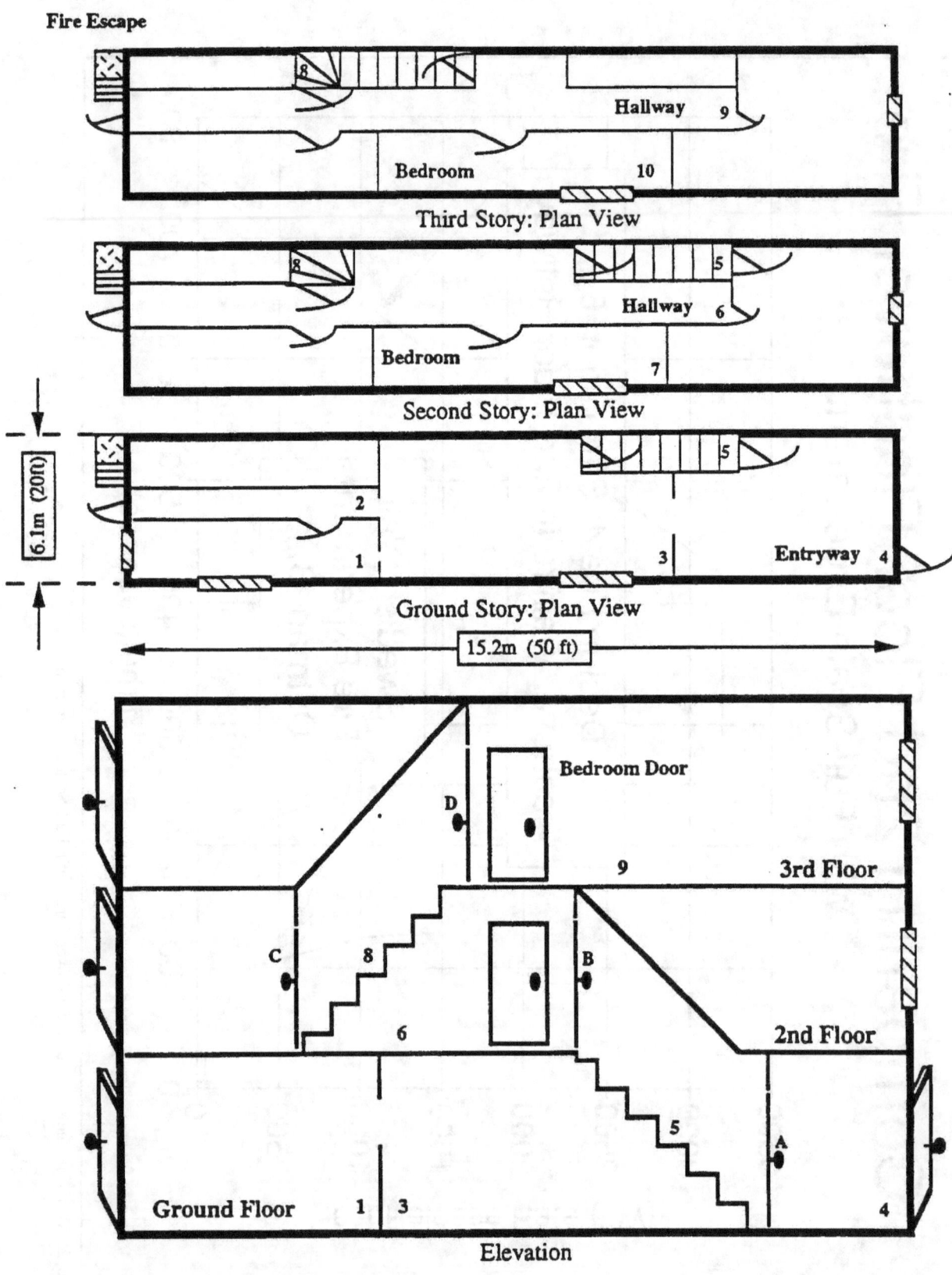

Figure 3. Plan & Elevation Views of a 3 Story Small Board and Care Home

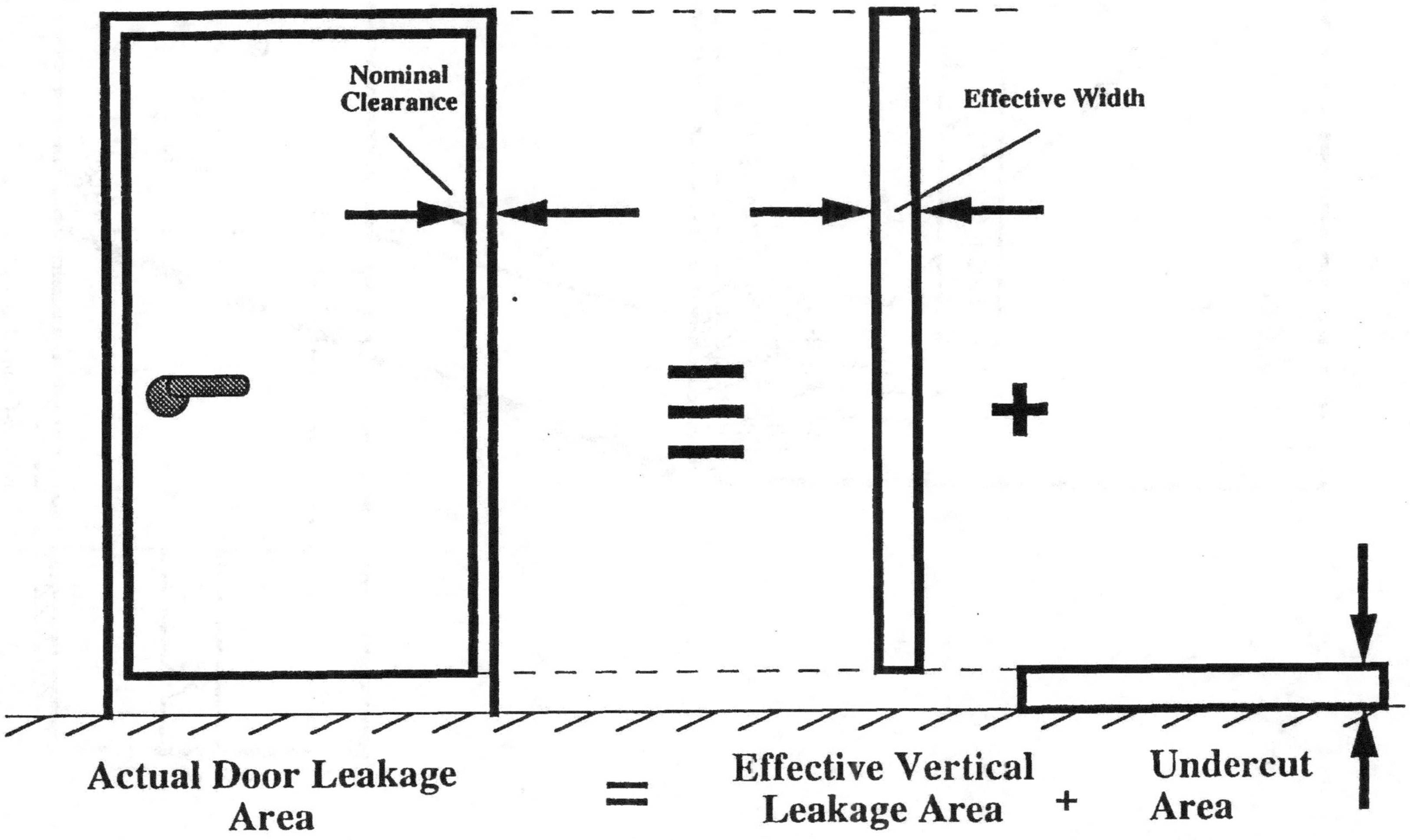

Effective Width

Nominal Clearance

+

=

Actual Door Leakage Area = **Effective Vertical Leakage Area** + **Undercut Area**

Figure 4. Method of Determining Equivalent Leakage Areas from Actual Door Leakage

33

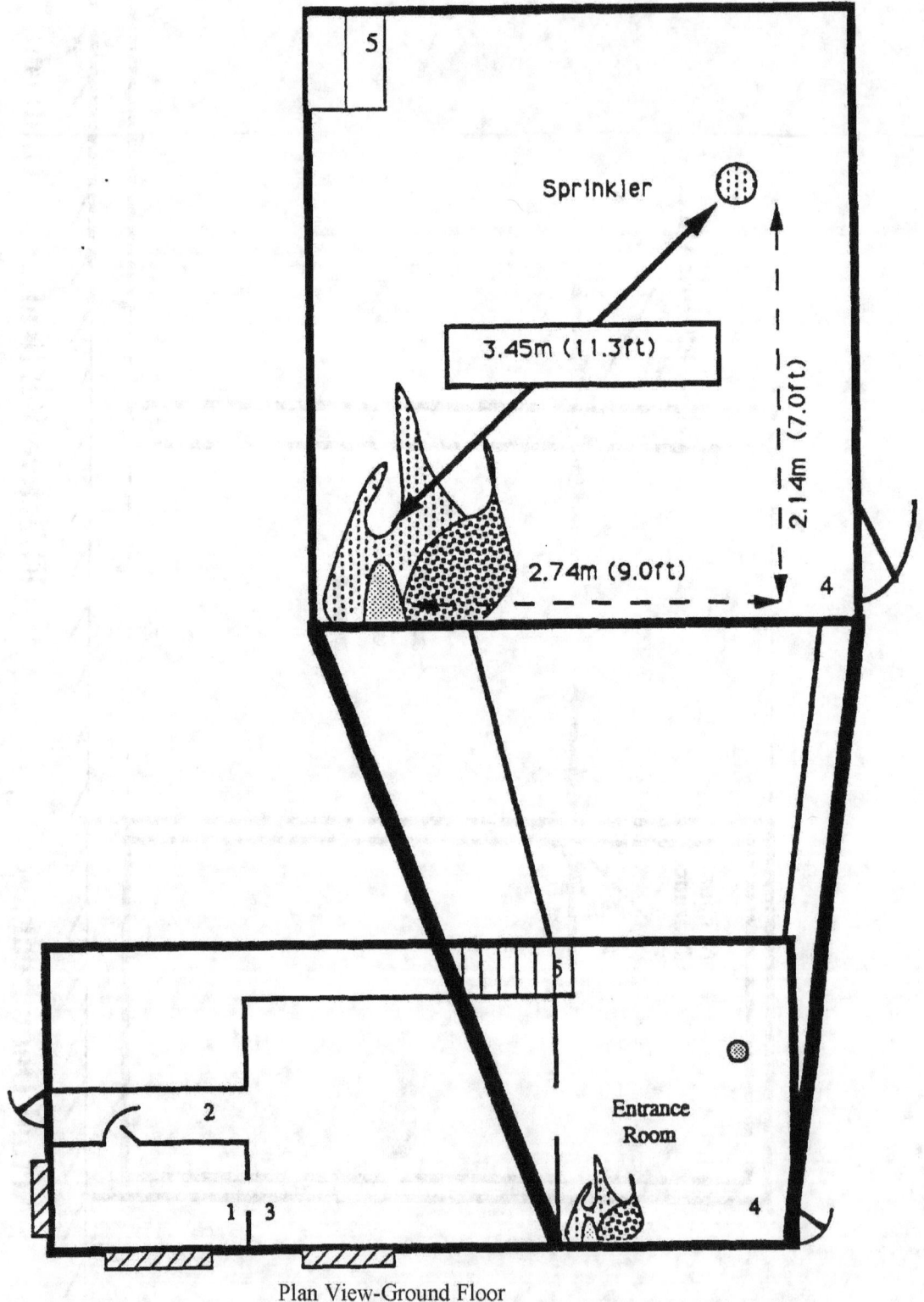

Figure 5. Sprinkler Location in the Ground Floor Entrance Room

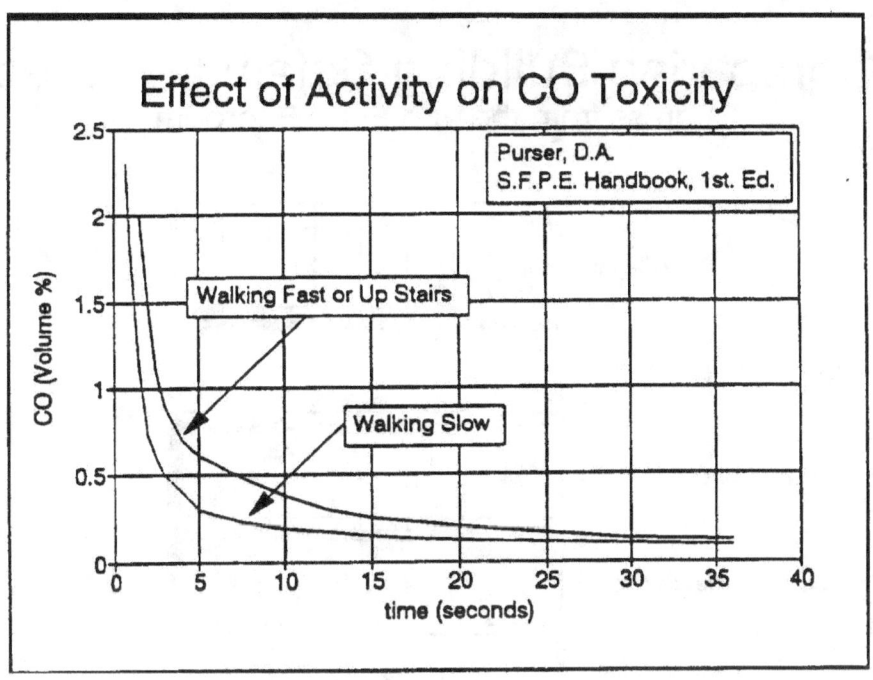

Figure 6. Effect of Physical Activity on Incapacitation during CO Inhalation

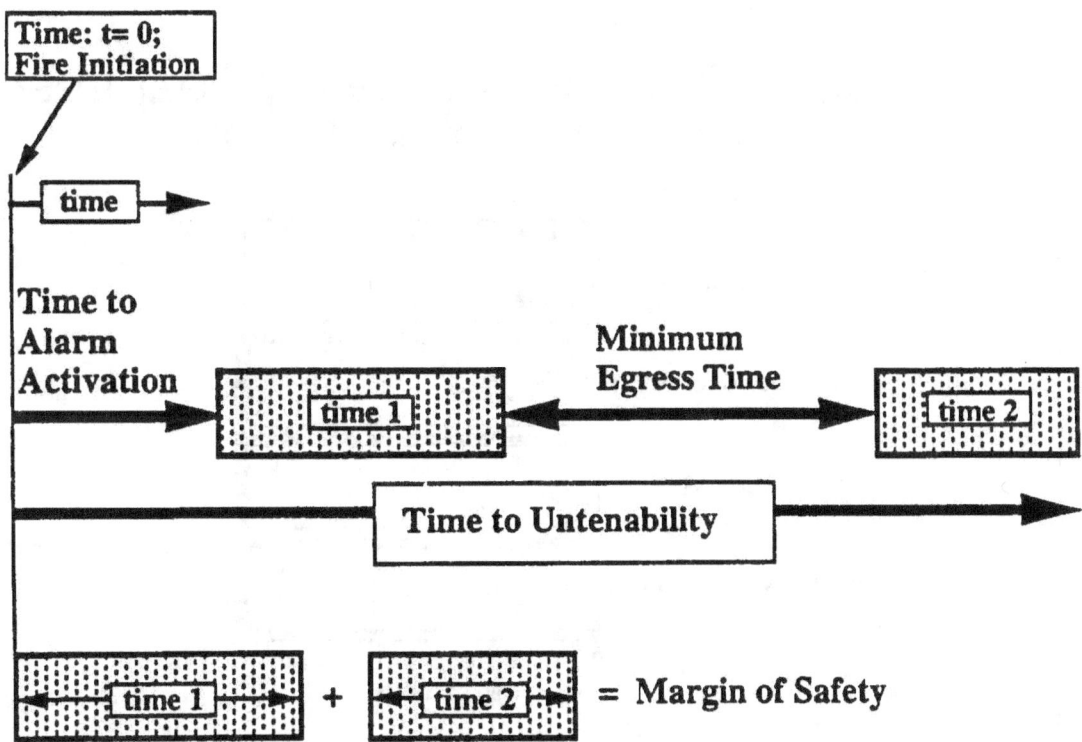

Figure 7. Margin of Safety Timeline

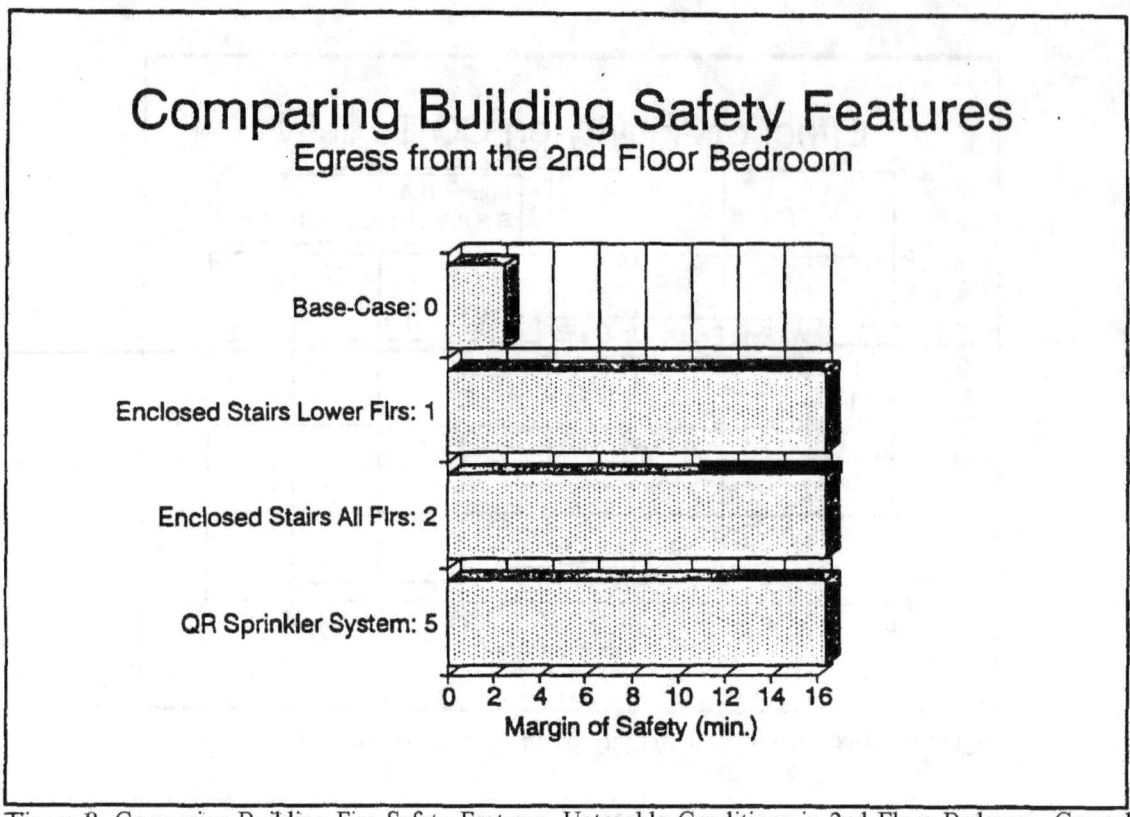

Figure 8. Comparing Building Fire Safety Features- Untenable Conditions in 2nd Floor Bedroom, Ground Floor Fire

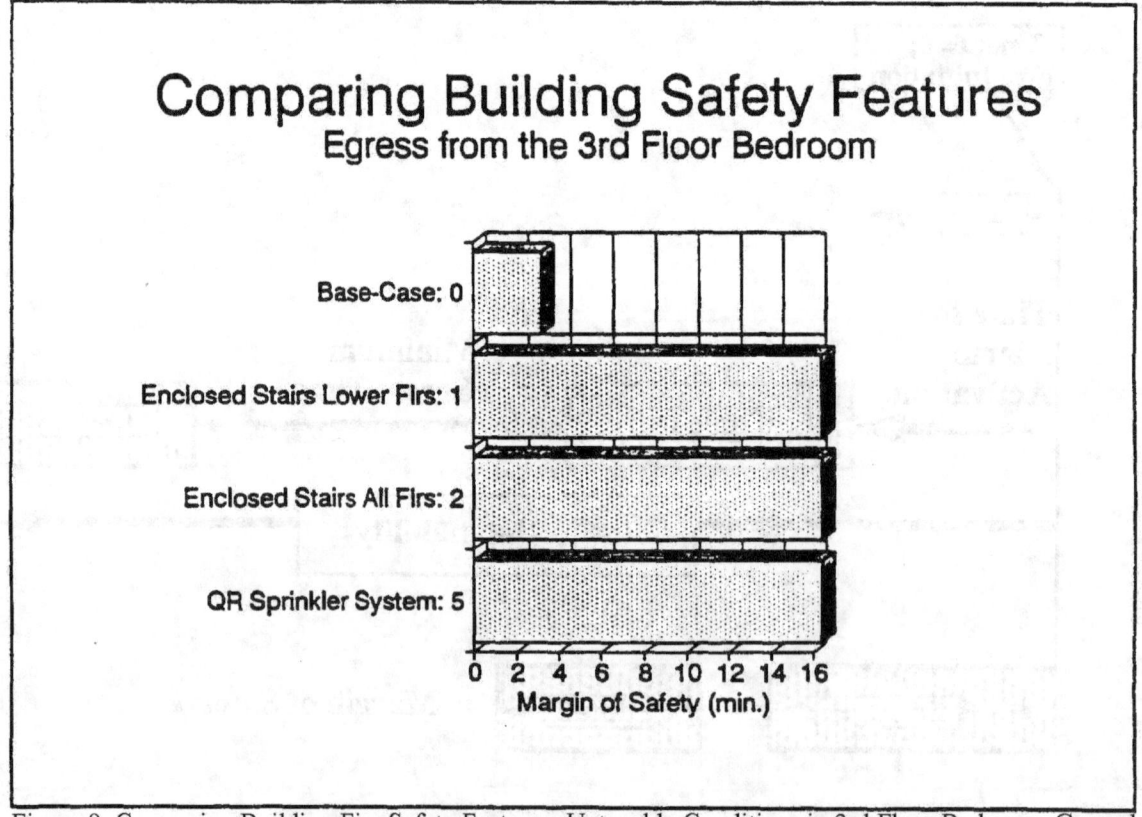

Figure 9. Comparing Building Fire Safety Features- Untenable Conditions in 3rd Floor Bedroom, Ground Floor Fire

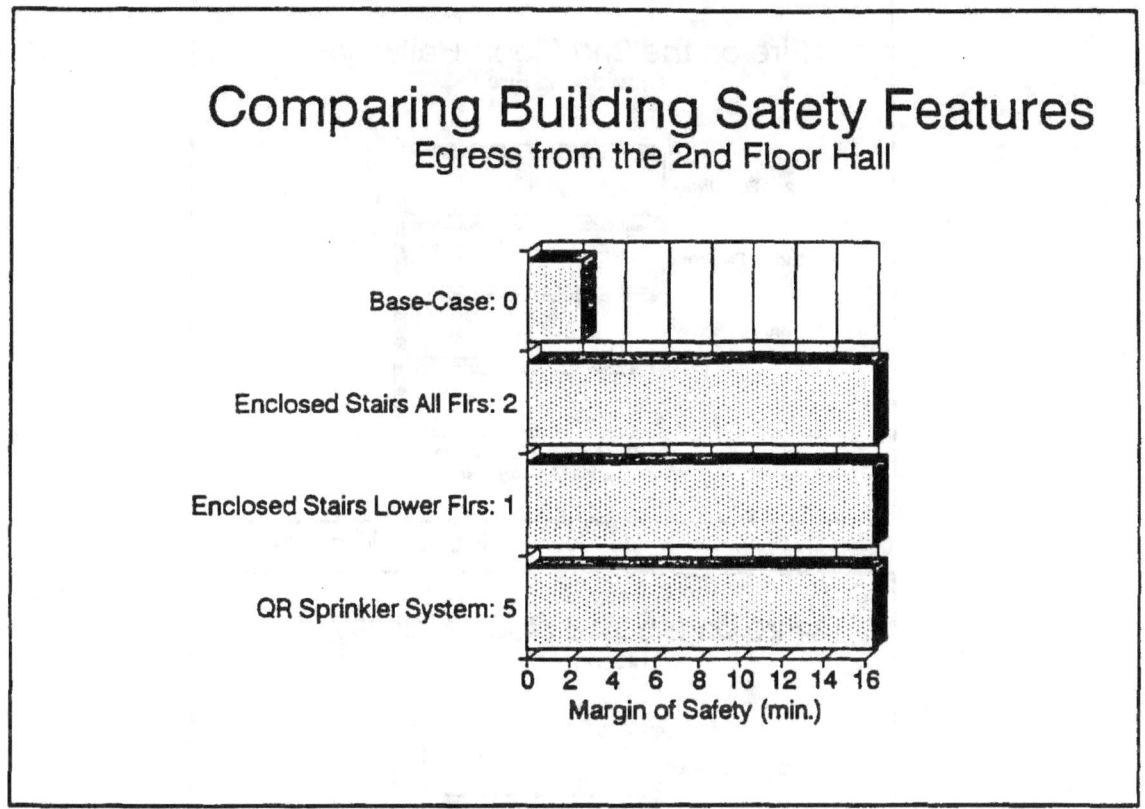

Figure 10. Comparing Building Fire Safety Features- Untenable Conditions in 2nd Floor Hallway, Ground Floor Fire

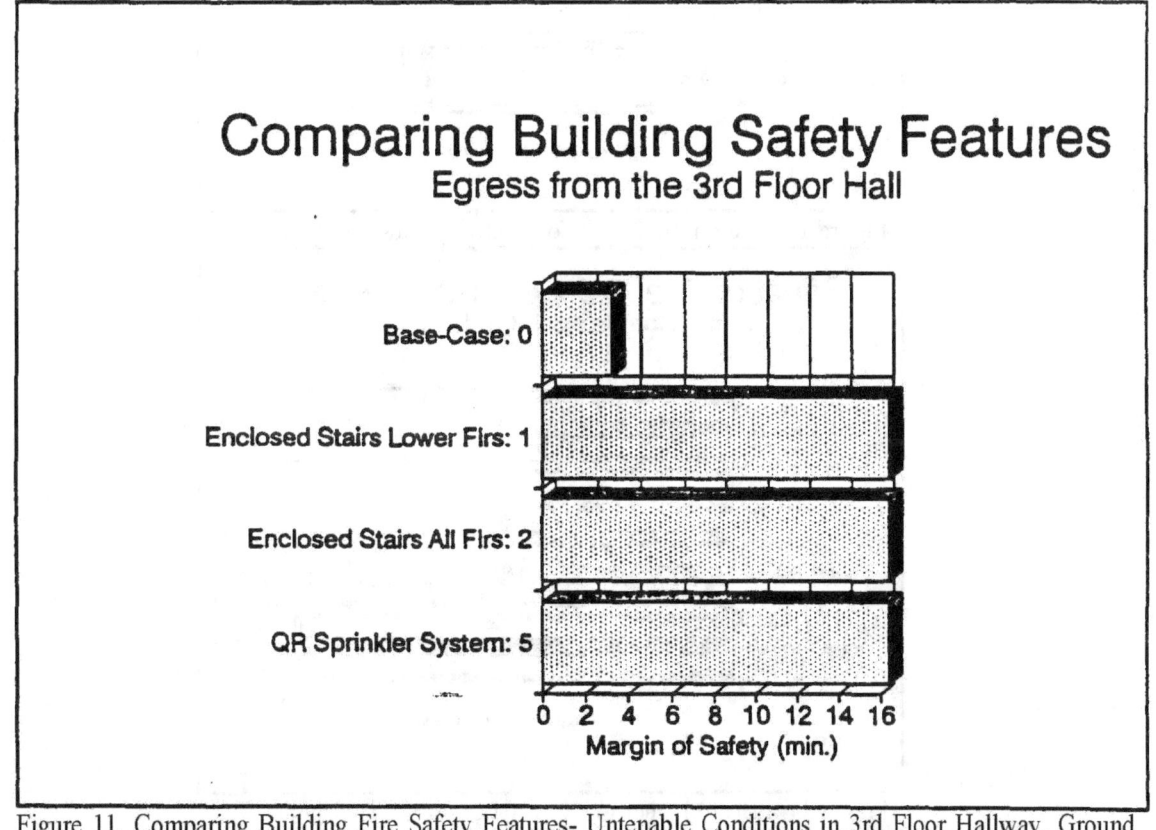

Figure 11. Comparing Building Fire Safety Features- Untenable Conditions in 3rd Floor Hallway, Ground floor Fire

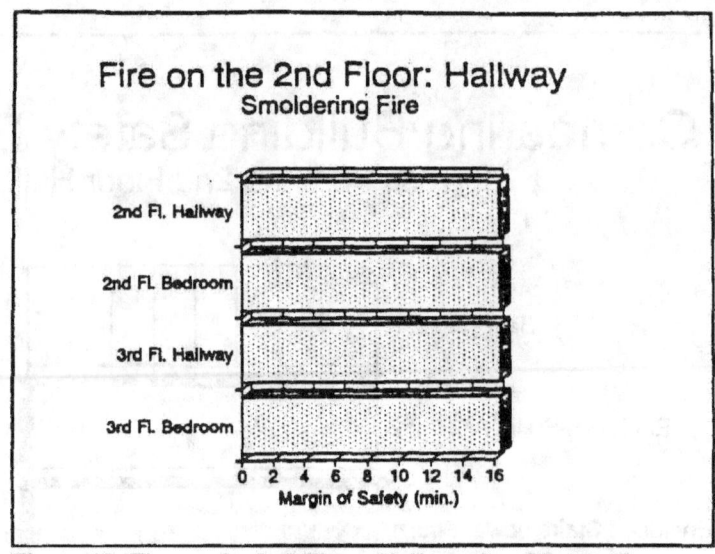

Figure 12. Fire on the 2nd Floor: Hallway-Smoldering Fire

Figure 13. Fire on the 2nd Floor. Hallway-Flashover Fire

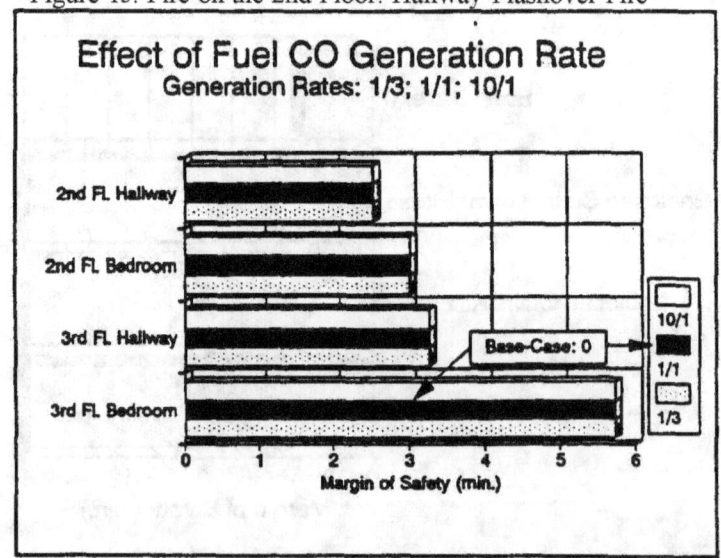

Figure 14. Effect of Fuel CO Generation Rates, Generation Rates of 1/3, 1/1, and 10/1 that of Base Case Generation Rates, Ground Floor Fire

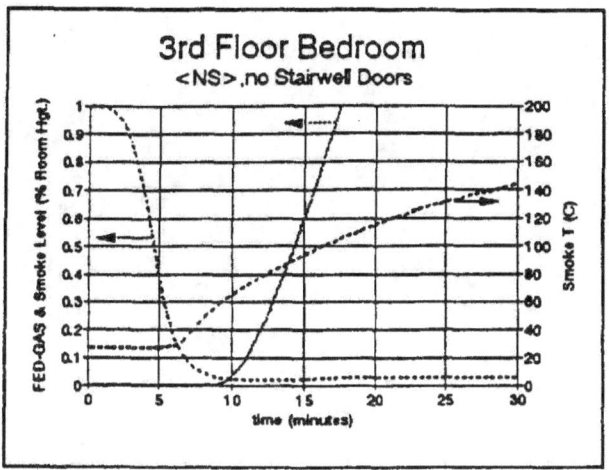

Figure 15a

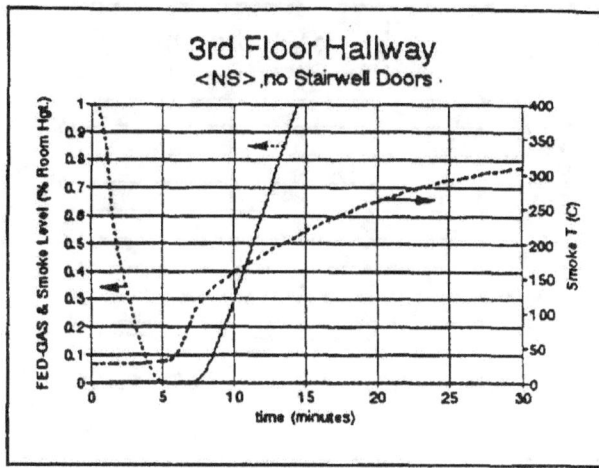

Figure 15b

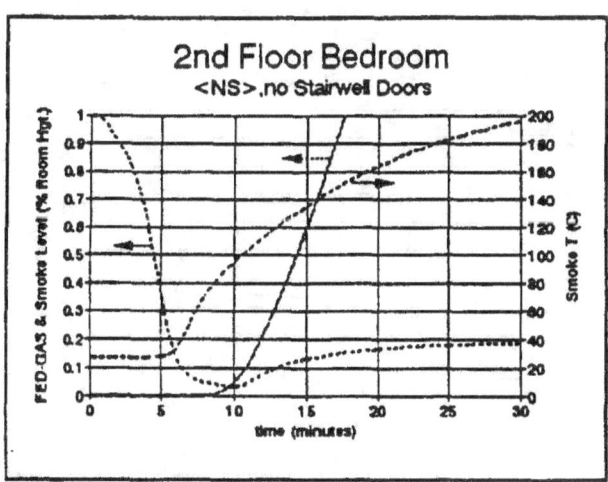

Figure 15c

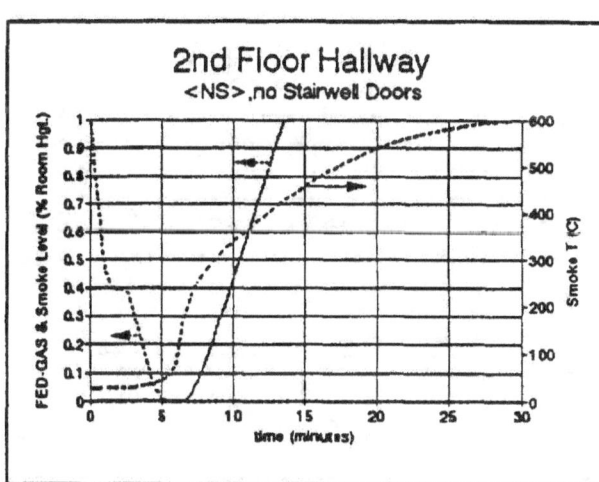

Figure 15d

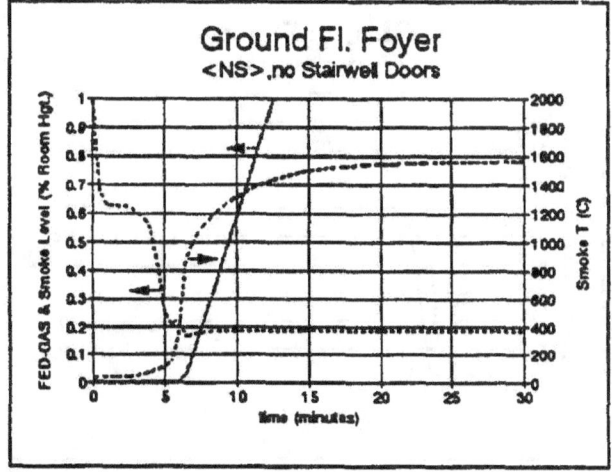

Figure 15e

Figure 15. Tenability Predictions: Nonsprinklered, No Stairwell Doors, Base Case

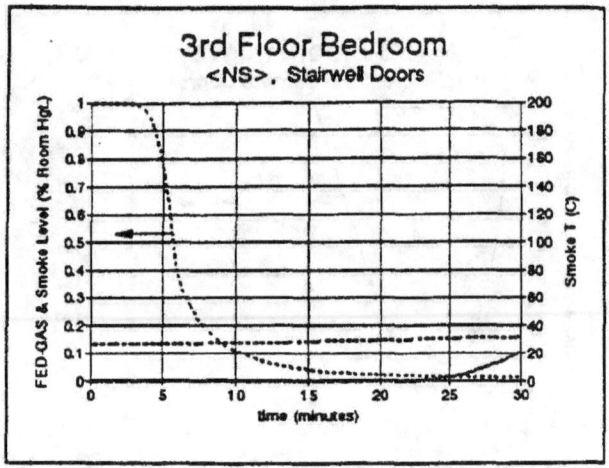

Figure 16a

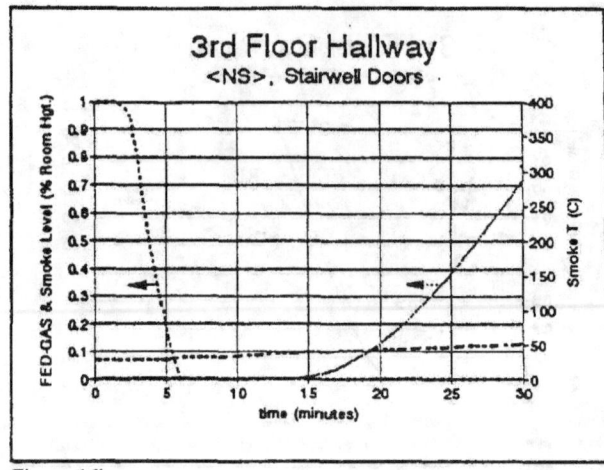

Figure 16b

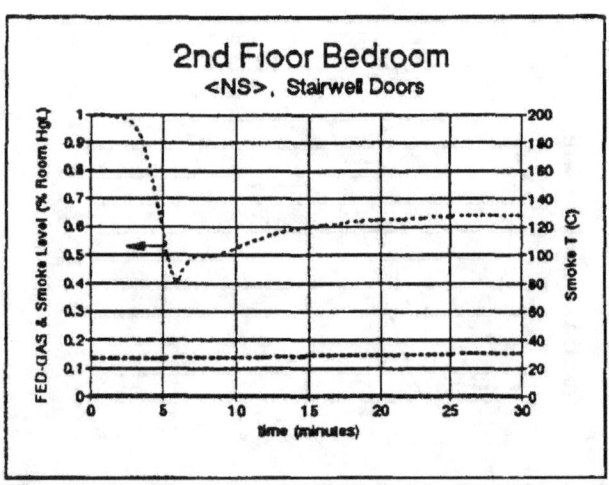

Figure 16c

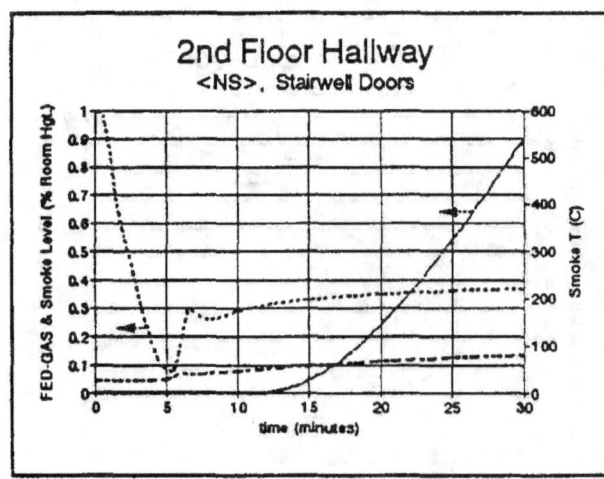

Figure 16d

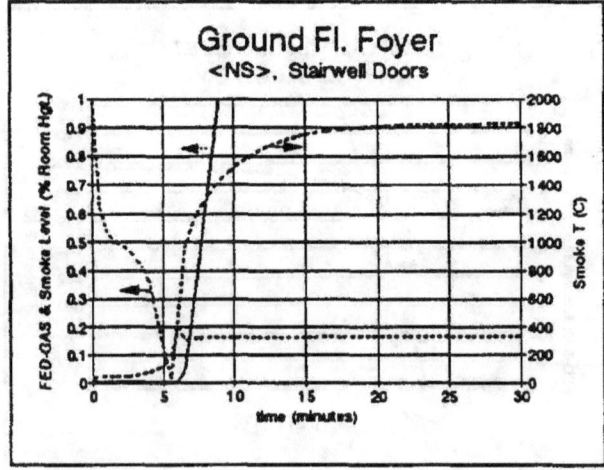

Figure 16e

Figure 16. Tenability Predictions: Nonsprinklered, Stairwell Doors per LSC

40

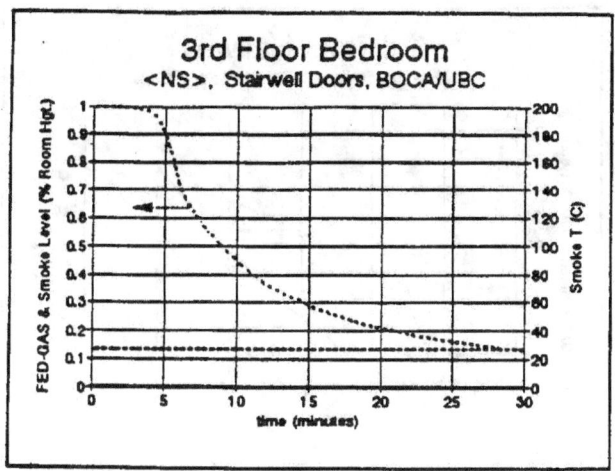

Figure 17a

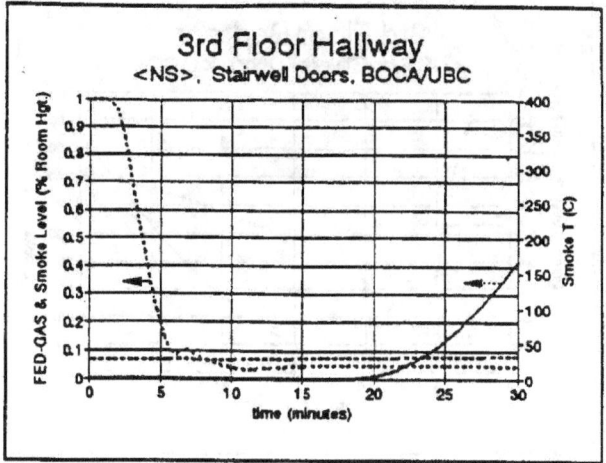

Figure 17b

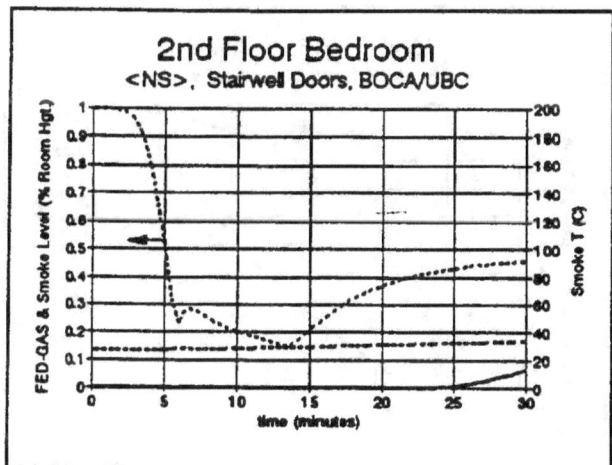

Figure 17c

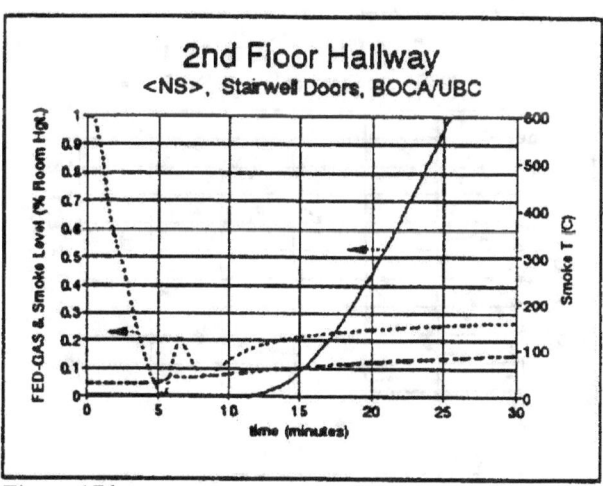

Figure 17d

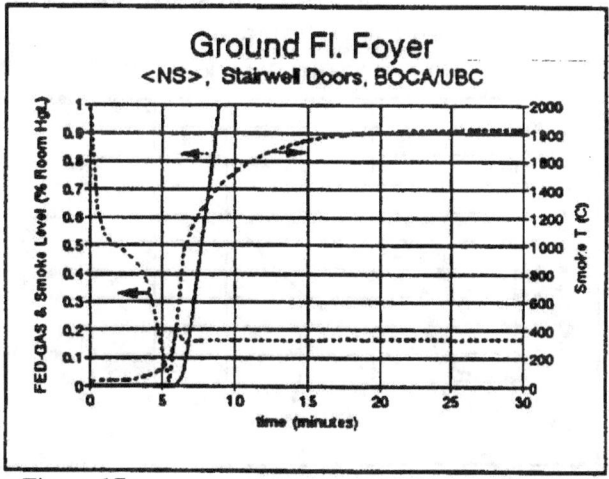

Figure 17e

Figure 17. Tenability Predictions: Nonsprinklered, Stairwell Doors per BOCA/UBC

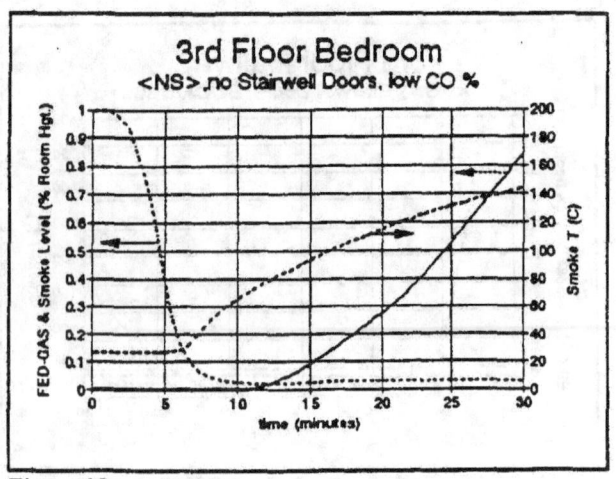

Figure 18a

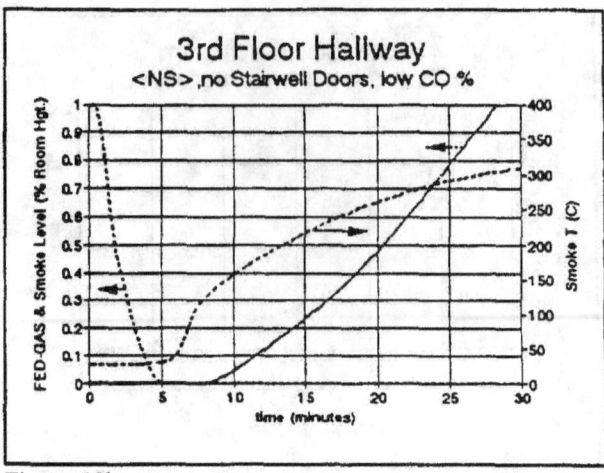

Figure 18b

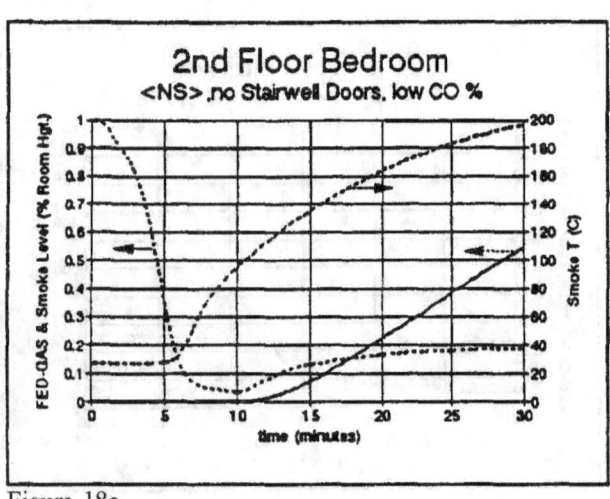

Figure 18c

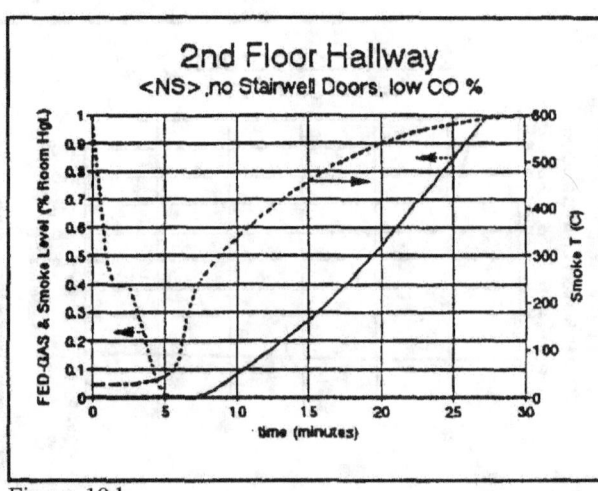

Figure 18d

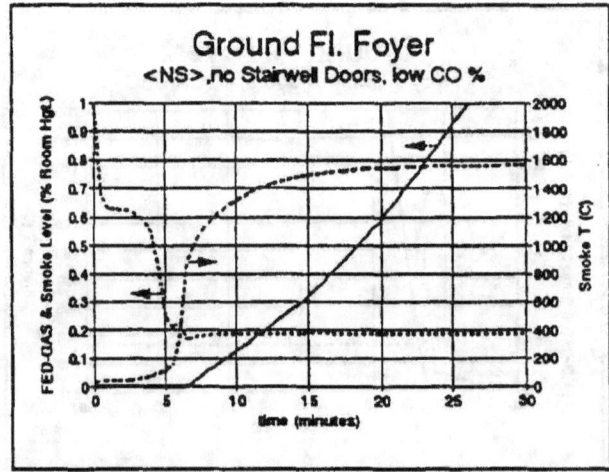

Figure 18e

Figure 18. Tenability Predictions: Base Case Fire: Low CO Production

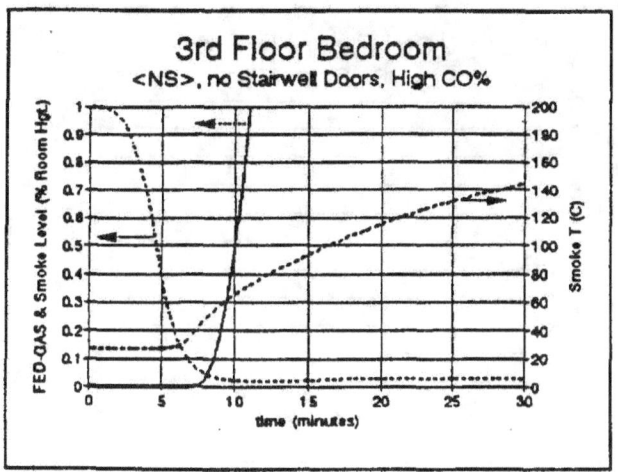

Figure 19a

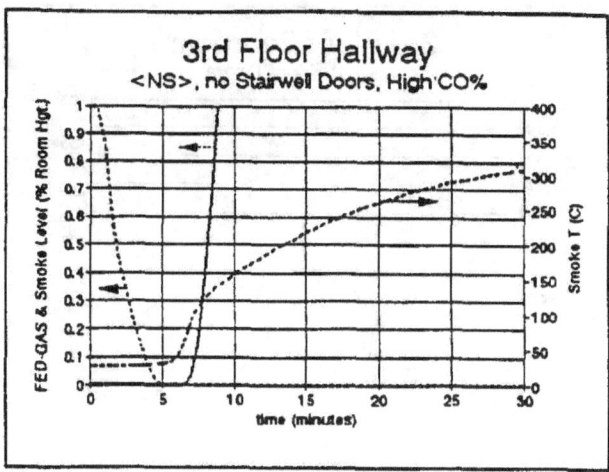

Figure 19b

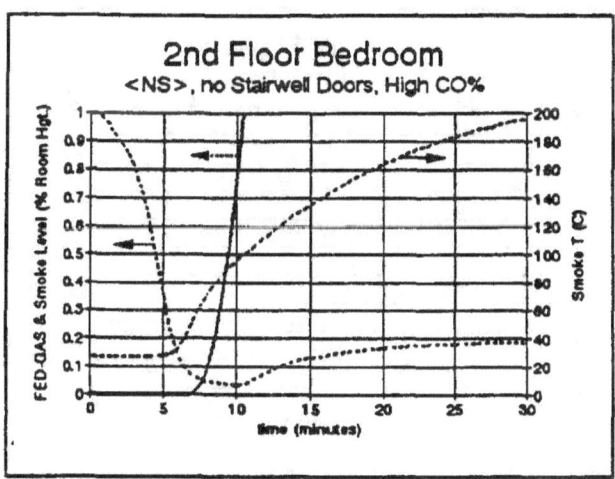

Figure 19c

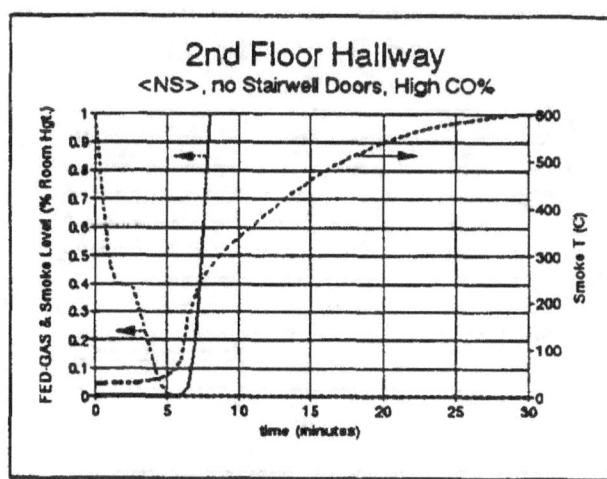

Figure 19d

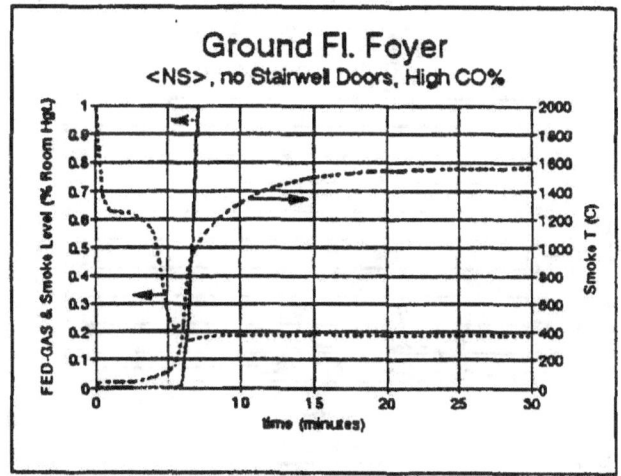

Figure 19e

Figure 19. Tenability Predictions: Base Case Fire: High CO Production

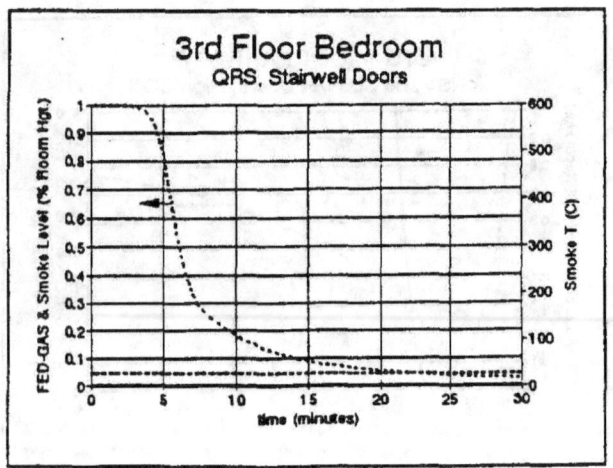

Figure 20a

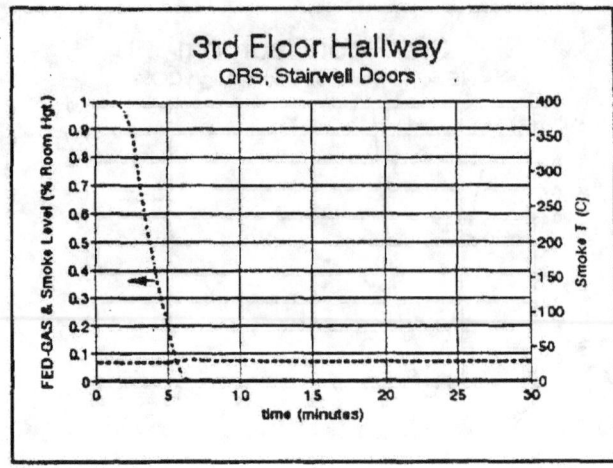

Figure 20b

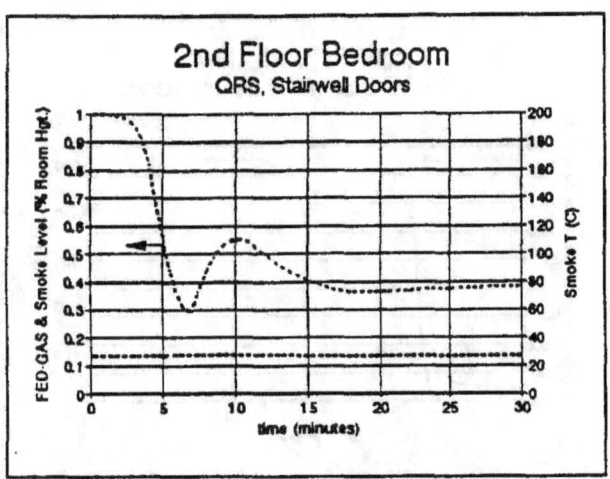

Figure 20c

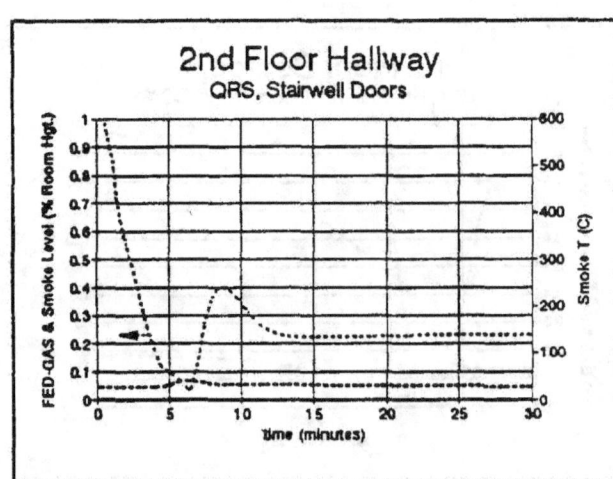

Figure 20d

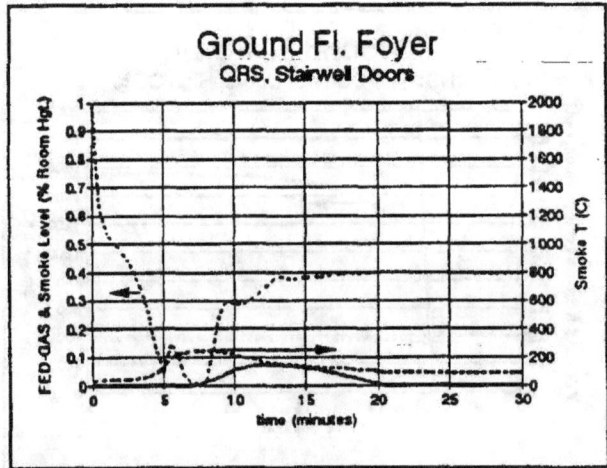

Figure 20e

Figure 20. Tenability Predictions: Base Case Fire, QR Sprinklers

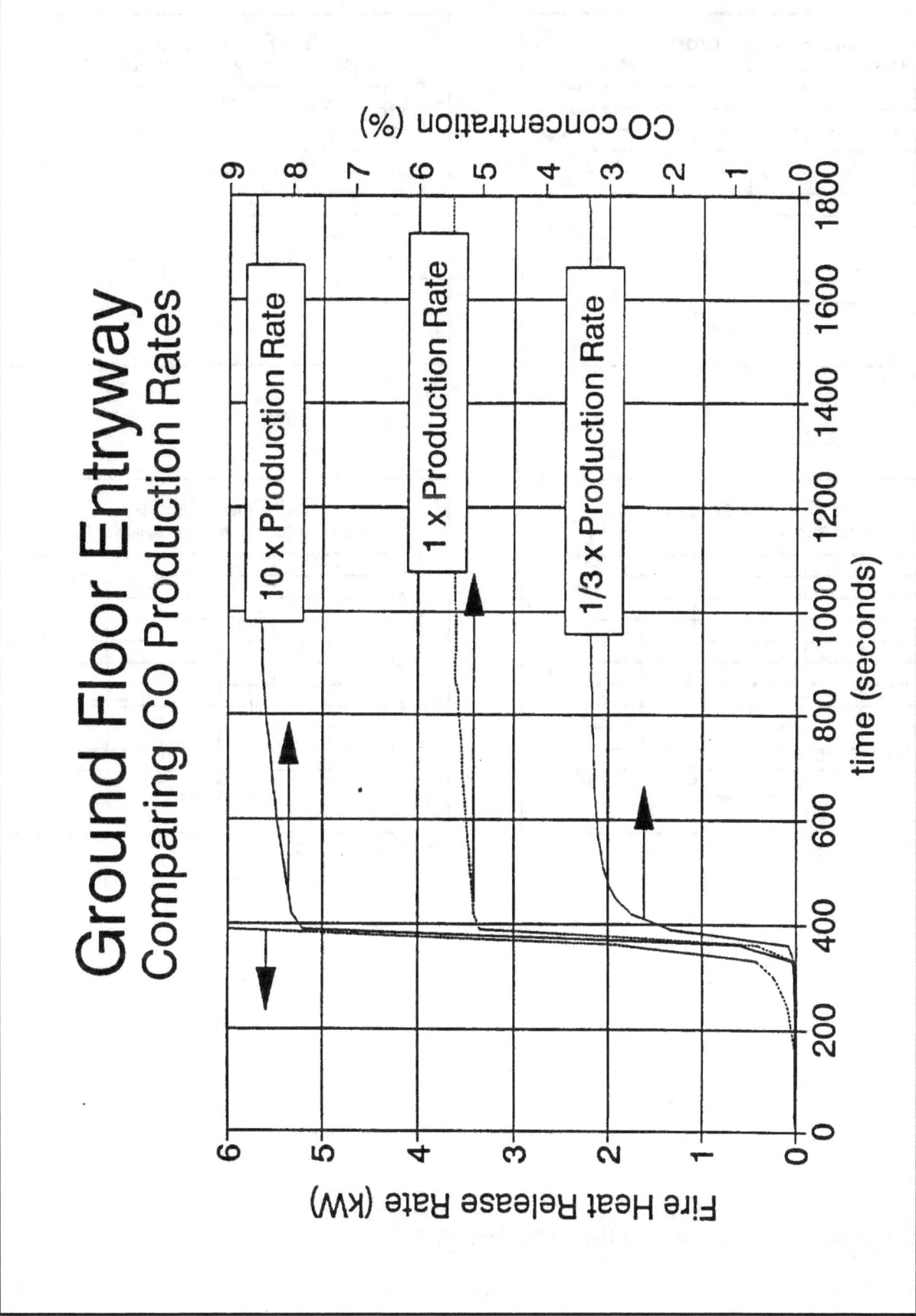

Figure 21. Comparing CO Generation Rates from Different Fires in the Ground Floor Entryway

45

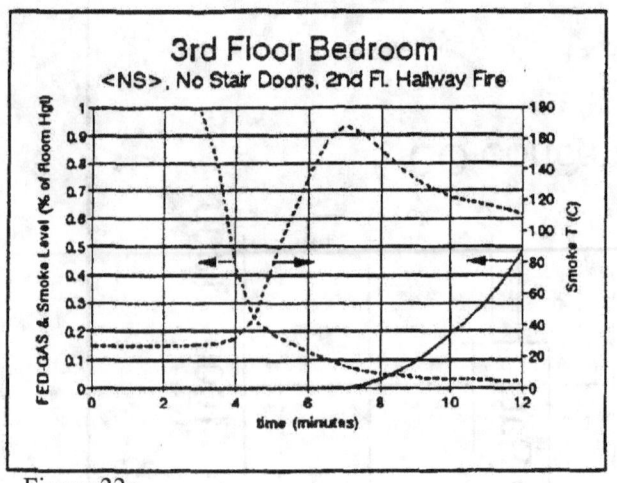

Figure 22a

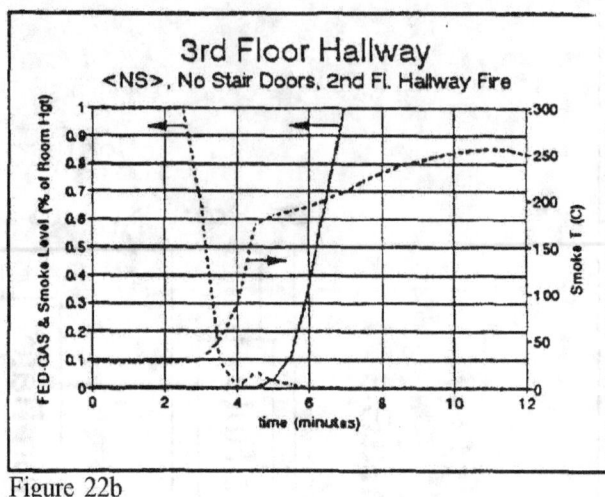

Figure 22b

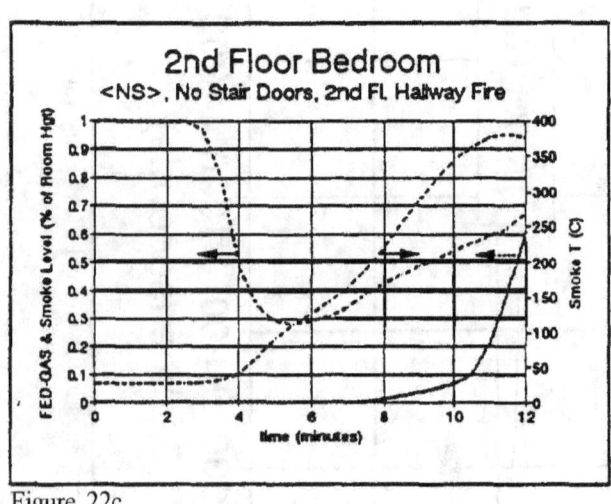

Figure 22c

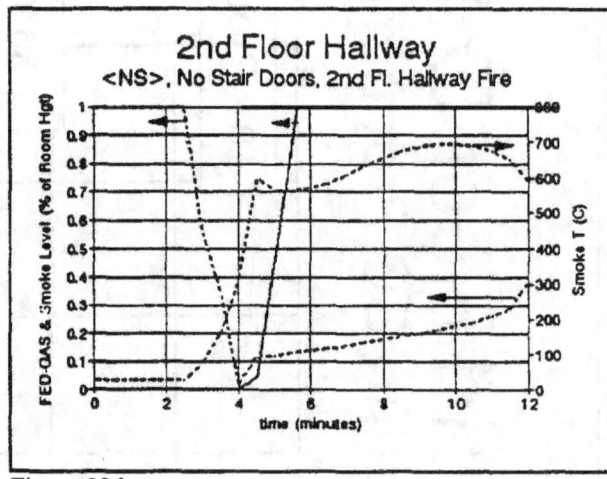

Figure 22d

Figure 22. Tenability Predictions: 2nd Floor Hallway Flashover Fire

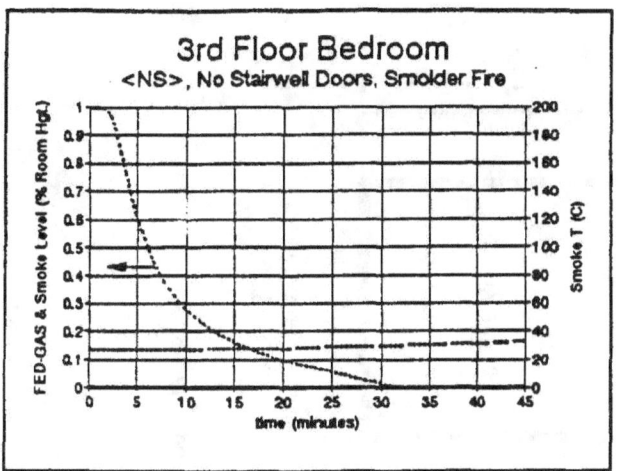

Figure 23a

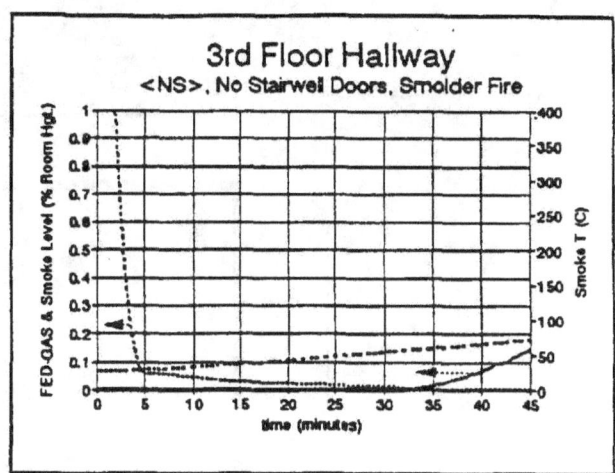

Figure 23b

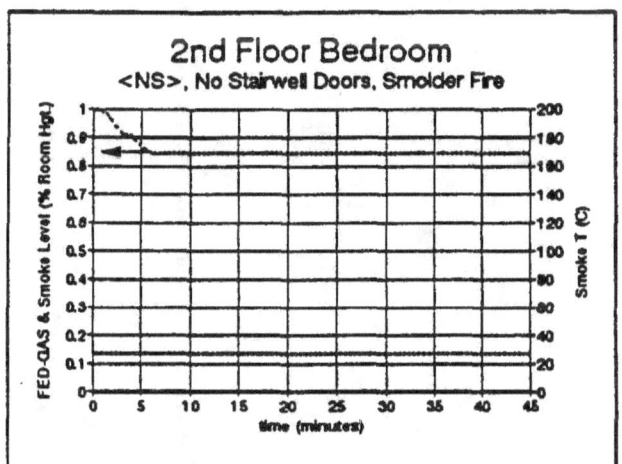

Figure 23c

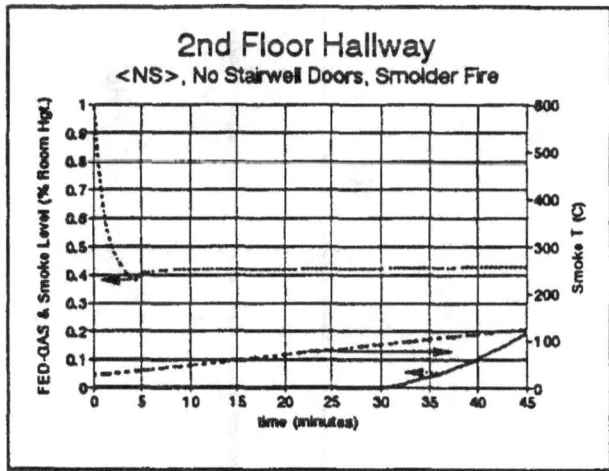

Figure 23d

Figure 23. Tenability Predictions: 2nd Floor Hallway Smoldering Fire

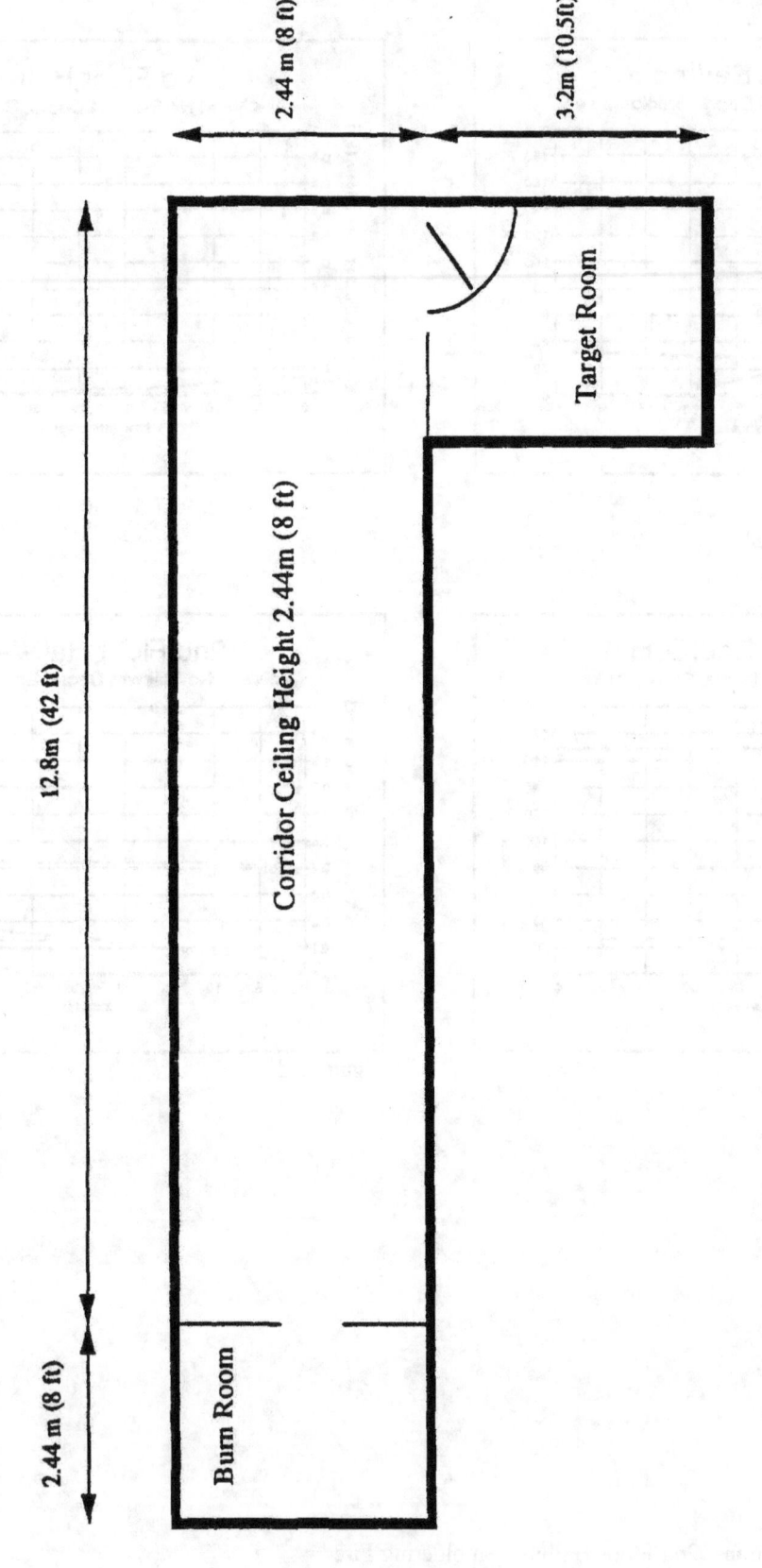

Figure 28. Plan View of Full-Scale Physical Layout of an Experimental Burn Room, Corridor and Target Room

10.0. Appendix A-Species Production Rates Used for Input to CFAST1.6 Fire Model

fire scenario	fuel	Pre-flashover CO/CO_2 Production rate (kg CO/kg CO_2)	Post-flashover CO/CO_2 Production Rate (kg CO/kg CO_2)	Pre-flashover Soot Production (kg C/kg CO_2)	Post-flashover Soot Production (kg C/kg CO_2)
Flashover Fire on Ground floor BaseCase	plastic and wood	.003	.067	.017	.020
Flashover Fire on Ground Floor with Low co Production Rate	plastic and wood	.003	.025	.017	.020
Flashover Fire on Ground Floor with High CO Production Rate	plastic and wood	-003	.67	-017	.020
Data source for the post-flashover production rates: [Brauskas 1991]. Data source for the pre-flashover production rates: DiNenno 1988]					

11.0 Appendix B-Sample CFAST1.6 Input Data File

```
VERSN    1 <NS>, lowr CO by factor 3
TIMES    1800   0    30    60    0
TAMB   300.  101300.      0.
EAMB   300.  101300.      0.
HI/F     0.00   0.00   0.00   0.00   0.00  2.74   2.74   2.74   5.33  5.33
WIDTH    4.32   4.32   4.81   4.95   3.16  10.71   3.61   0.71   7.85  5.18
DEPTH    2.18   2.50   4.46   4.95   1.02   1.52   5.46   1.02   1.42  4.17
NIGH     2.44   2.74   2.74   2.74    3.13  2.59   2.59   4.82   2.23  2.23
HVENT   1  2  1    0.760    2.032    0.000
HVENT   1  3  1    0.560    2.240    1.450
HVENT   1 11  1    1.680    2.310    1.020 0.0
HVENT   2  3  1    0.860    2.130    0.000
HWENT   3  4  1    1.520    2.130    0.000
HVENT   3 11  1    1.83     2.130    0.61 0.0
HVENT   4  5  1    0.813    2.740    0.000
HVENT   4 11  1    0.860    2.032    0.000 0.000
HVENT   5  6  1    0.813    5.330    2.740
HEVENT  6  7  1    0.023    2.032    0.019
HVENT   6  7  2    0.762    0.019    0.000
HVENT   6  8  1    0.813    2.032    0.000
HVENT   6 11  1    0.018    2.590    0.000 0.000
HVENT   7 11  1    0.61     1.140    0.610 0.000
HVENT   8  9  1    0.813    4.820    2.590
HVENT   9 10  1    0.023    2.032    0.019
HVENT   9 10  2    0.762    0.019    0.000
HVENT   9 11  1    0.018    2.230    0.000 0.000
HVENT  10 11  1    0.360    1.020    0.610 0.000
CVENT   1  2  1    0.00   0.00   0.00  1.00  1.00   1.00   1.00   1.00   1.00   1.00   1.00   1.00   1.00
CVENT   1  3  1    1.00   1.00   1.00  1.00  1.00   1.00   1.00   1.00   1.00   1.00   1.00   1.00   1.00
CVENT   1 11  1    0.00   0.00   0.00  0.00  0.00   0.00   0.00   0.00   0.00   0.00   1.00   1.00   1.00
CVENT   2  3  1    0.00   0.00   0.00  0.00  0.00   0.00   0.00   0.00   1.00   1.00   1.00   1.00   1.00
CVENT   3  4  1    1.00   1.00   1.00  1.00  1.00   1.00   1.00   1.00   1.00   1.00   1.00   1.00   1.00
CVENT   3 11  1    0.00   0.00   0.00  0.00  0.00   0.00   0.00   0.00   0.00   0.00   1.00   1.00   1.00
CVENT   4  5  1    1.00   1.00   1.00  1.00  1.00   1.00   1.00   1.00   1.00   1.00   1.00   1.00   1.00
CVENT   4 11  1    0.00   0.00   0.00  0.00  0.00   0.00   0.00   0.00   0.00   0.00   1.00   1.00   1.00
CVENT   5  6  1    1.00   1.00   1.00  1.00  1.00   1.00   1.00   1.00   1.00   1.00   1.00   1.00   1.00
CVENT   6  7  1    1.00   1.00   1.00  1.00  1.00   1.00   1.00   1.00   1.00  1.00   1.00   1.00   1.00
CVENT   6  7  2    1.00   1.00   1.00  1.00  1.00   1.00   1.00   1.00   1.00  1.00   1.00   1.00   1.00
CVENT   6  8  1    1.00   1.00   1.00  1.00  1.00   1.00   1.00          1.00   1.00   1.00   1.00   1.00
CVENT   6 11  1    1.00   1.00   1.00  1.00  1.00   1.00   1.00   1.00   1.00   1.00   1.00   1.00   1.00
CVENT   7 11  1    1.00   1.00   1.00  1.00  1.00   1.00   1.00   1.00  1.00   1.00   1.00   1.00   1.00
CVENT   8  9  1    1.00   1.00   1.00  1.00  1.00   1.00   1.00   1.00  1.00   1.00   1.00   1.00   1.00
CVENT   9 10  1    1.00   1.00   1.00  1.00  1.00   1.00   1.00         1.00  1.00   1.00   1.00   1.00
CVENT   9 10  2    1.00   1.00   1.00  1.00  1.00   1.00   1.00          1.00   1.00   1.00   1.00   1.00
CVENT   9 11  1    1.00   1.00   1.00  1.00  1.00   1.00   1.00   1.00   1.00   1.00   1.00   1.00   1.00
CVENT  10 11  1    1.00   1.00   1.00  1.00  1.00   1.00   1.00   1.00   1.00   1.00   1.00   1.00   1.00
CEILI GYP1/2   GYP1/2   GYP1/2   GYP1/2   GYP1/2   GYP1/2              GYP1/2   GYP1/2   GYP1/2
WALLS GYP1/2   GYP1/2   GYP1/2   GYP1/2   GYP1/2   GYP1/2   GYP1/2   GYP1/2   GYP1/2   GYP1/2
CHEMI   16.    0.    1.0    19000000.   300.400.     0.
LFBO    4
LFBT    2
FPOS     0.20  1.00  0.00
FTIME     1.  152.  153.  195.  200.   250.  300.  325.  350.  360.  361.  1000.
FMASS  0.0000 0.0006 0.0006 0.0009 0.0021 0.0024 0.0050 0.0121 0.0185 0.0355 0.1015 0.3158 0.3189
FHIGH    0.00   0.00   0.00   0.00   0.00   0.00   0.00   0.00   0.00   0.00   0.00   0.00   0.00
FAREA    0.00   0.00   0.00   0.00   0.00   0.00   0.00   0.00   0.00   0.00   0.00   0.00   0.00
FQDOT 0.00 1.14E+04 1.14E+04 1.71E+04 3.99E+04 4.56E+04 9.50E+04 2.30E+05 3.51E+05 6.74E+05 1.93E+06 6.00E+06 6.00E+06
WET OFF
HCR    0.130 0.130 0.130 0.130 0.130 0.130 0.130 0.130 0.130 0.130 0.130 0.130 0.130
CO     0.003 0.003 0.003 0.003 0.003 0.003 0.003 0.003 0.003 0.003 0.025 0.025 0.025
OO     0.017 0.017 0.017 0.017 0.017 0.017 0.017 0.017 0.017 0.017 0.020 0.020 0.020
DUMPR STBGOCN5.DMP
```

12.0. Appendix C- Permissible Door Leakage Areas per Building/Fire Safety Codes

Door Description	Location of Leakage Area	Clearance mm (in.)	Leakage Area $m^2 \cdot 10^3$ (in.^{2})	Verticle Leakage Effective Width ($W_{effective}$) mm (in.)
Bedroom Door: Nominal Clearance 30 in. Wide x 80 in. High NFPA 252 ASIM E152-181	top of door	3.18 (.12.5)	2.42 (3.75)	
	latch side of door	3.18 (.125)	6.41 (9.94)	
	hinge eide of door	3.18 (.125)	6.41 (9.94)	7.55 (.297)
	undercut of door	9 53 (-375)	7.26 (11.3)	
Bedroom Door: Maximum Latch Deflection 30 in. Wide x 80 in. High NFPA 252 ASTM E152-18a	top of door	3.18 (.125)	2.42 (3.75)	
	latch ride of door	12.7 (J)	25.7 (39.8)	
	hinge ride of door	3.18 (.125)	6.41 (9.94)	17.1 (.672)
	undercut of door	9.53 (375)	7.26 (11.3)	
Bedroom Door: Norminal Clumocu 30 in. Wide x 80 in. High NFPA 80	top of door	9.53 (375)	7.26 (11.3)	
	latch side of door	9.53 (.375)	19.1 (29.6)	
	hinge side of door	953 (375)	19.1 (29.6)	22.7 (.893)
	undercut of door	19.1 (.75)	14.5 (22.5)	

In the CFAST1.6 mathematical computer fire model, the leakage area of an actual door was represented by two equivalent vent areas. These leakage areas represented areas through which smoke could flow past a door. The first leakage area represented the crack formed by the undercut of the door; the second leakage area represented the cracks formed everywhere above the undercut. This second leakage area is called the effective vertical leakage area because it is effectively simplified the representation of these actual leakage areas. The area of the effective vertical leakage is determined by adding all leakages areas found above the undercut and dividing by the height of the door (minus the height of the undercut).

Doors protecting stairways were 0.813 m (32 in.) wide while the doors listed in the above table are 0.762 m (30 in.) wide. The doors in the above table were sleeping rooms doors which opened onto the hallway. Although leakage area numbers for the stairwell doors are not presented, the calculational approach is similar to that presented above.

13.0. Appendix D–Evacuation Travel Times

Typical movement speeds for adults are 1.27 m/s (250 feet/min) on level surfaces and 0.2 m/s (40 vertical feet/min) on stairs wider than 0.81 m (32 in.). Stair pitch influences vertical travel speed and the 0.2 m/s travel rate presumes the stair riser and tread dimensions are 0.178 m x 0.279 m (7 in. x 11 in.). Travel speeds on stairs are also influenced by width. In cases where stairs are narrower than 0.813 m (32 in.) and/or steeper than the 7 x 11 inch stairs, actual travel speeds should be adjusted. The **EGRESS** module within **FPEtool3.0** accounts for these effects.

For simulating evacuation movement of disabled residents the Americans with Disabilities Act [DoJ 1991] provides some guidelines, although the accuracy of these guidelines has not been substantiated. For disabled individuals travelling on horizontal surfaces, the ADA recommends 0.47 m/s (90 fWmin). The ADA further states that no vertical travel on stairs may be considered (ramps are allowed) and a 2 minute rest for every 100 feet of travel is required.

It is possible to use the ADA recommendations to glimpse at how the margin of safety may be affected when individuals with mobility limitations are expected to self-evacuate. One must first ignore the assumption that the disabled will not use the stairs. Given this assumption, the margins of safety for disabled individuals may be calculated from the additional travel time needed by the disabled residents as cited in the following tables. B&C occupants with mobility limitations should be receive regular evacuation training and evaluation to ensure their performance remains effective.

Table D-I

					Fast		Slow	
Point	Horizontal Distance m (ft)	Vertical Height m (ft)	Stairwell Width m (ft)	Doorway Width m (ft)	Travel Speed m/s (ft/min)	Travel time (sec)	Travel Speed m/s (ft/min)	Travel time (sec)
A-B	3.61 (11.8)				1.27 (250)	2.8	0.47 (90)	6.7
B-C			.762 (2.5)		1 Door/sec	1	1 Door/sec	1.
C-D	2.03 (6.67)				1.27 (250)	1.6	0.47 (90)	4.3
D-E'	5 (16.4)				1.27 (250)	3.9	0.47 (90)	10.6
E'-F'				.813 (2.67)	1 Door/sec	1.	0.47 (90)	1.
Total						10.30		23.60

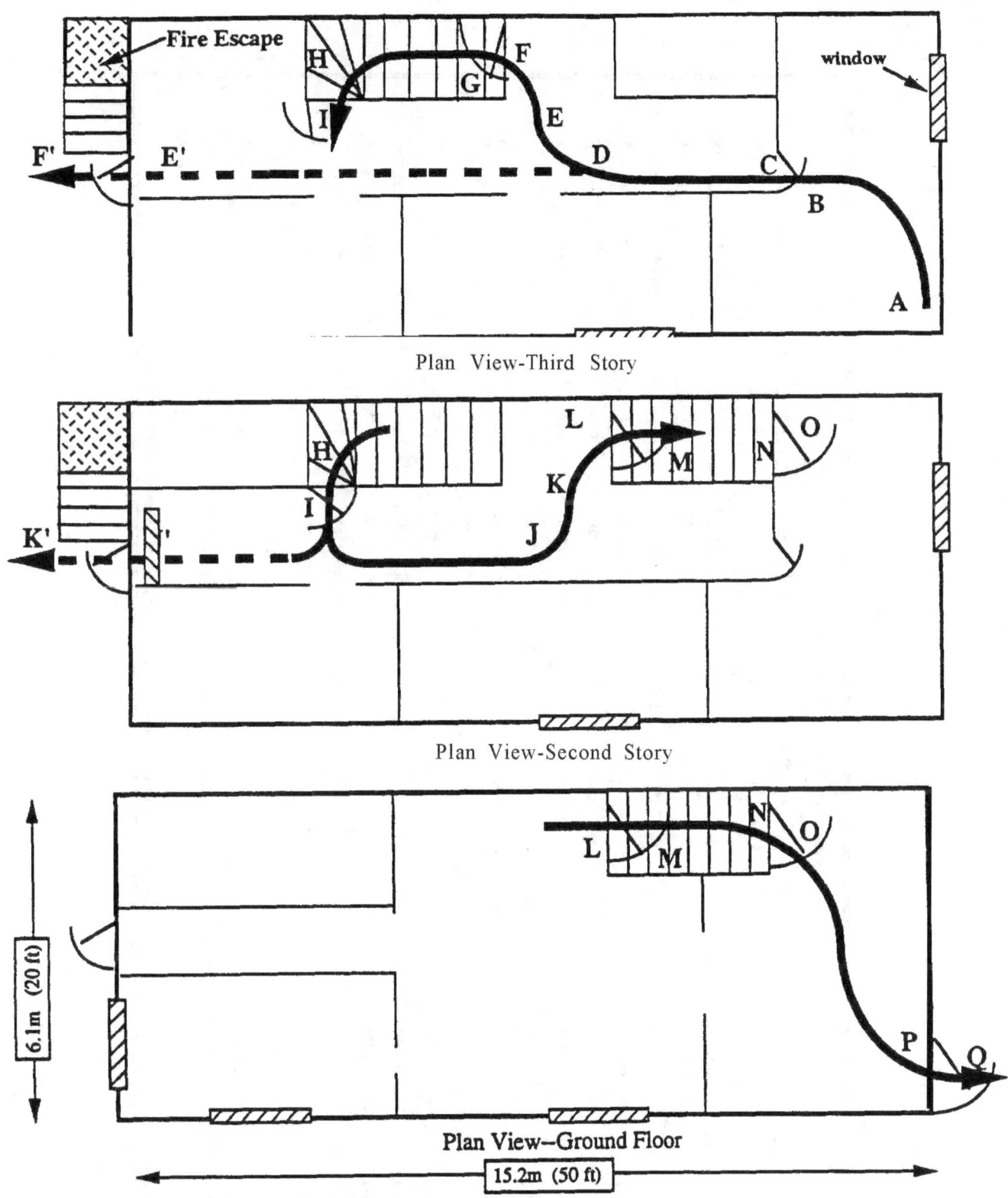

Plan View-Third Story

Plan View-Second Story

Plan View-Ground Floor

Figure D-l. Egress Paths from the Third Story of a Small Board and Care Home

Table D-2

Evacuation Movement Calculations: Base Case Scenario
3 Story Home, 1 Stairwell without Door Enclosures
Movement Fran 3rd Story Bedroom to Ground Floor Exterior Doorway

Point	Horizontal Distance m (ft)	Vertical Height m (ft)	Stairwell Width m (ft)	Doorway Width m (ft)	Fast Travel Speed m/s (ft/min)	Travel the (sec)	Slow Travel Speed m/S (ft/min)	Travel time (set)
A-B	3.61 (11.8)				1.27 (250)	2.8	0.47 (90)	6.7
B-C				.762 (2.5)	1 Door/sec	1	1 Door/sec	1.
C-D	2.03 (6.67)				1.27 (250)	1.6	0.47 (90)	4.3
D-E	.787 (2.58)				1.27 (250)	0.6	0.47 (90)	1.7
E-F	.787 (2.58)				1.27 (250)	0.6	0.47 (90)	1.7
F-G				.813 (2.67)	1 Door/sec	1.	1 Door/sec	1.
G-H		2.59 (8.50)	.813 (2.67)		0.19^1 (40)	13.6	0.07 (16)	37.0
H-I				.813 (2.67)	1 Door/sec	1.	0.47 (90)	1.
I-J	3.16 (10.4)				1.27 (250)	2.5	0.47 (90)	6.7
J-K	.787 (2.58)				1.27 (250)	.6	0.47 (90)	1.7
K-L	.787 (2.58)				1.27 (250)	.6	0.47 (90)	1.7
L-M				.813 (2.67)	1 Door/sec	1.	1 Door/sec	1.
M-N		2.74 (9.00)	.813 (2.67)		0.19^1 (40)	1	0.08 (16)	39.1
N-O				.813 (2.67)	1 Door/sec	1.	1 Door/sec	1.
O-P	7.00 (23.0)				1.27 (250)	5.5	0.47 (90)	14.9
P-Q				.813 (2.67)	1 Door/sec	1.	1 Door sec	1.
Total						46.50		112.10

'travel speed is reduced 6.5% from the 0.2 m/s rate to account for stairs with 8" and not 7" risers.

54

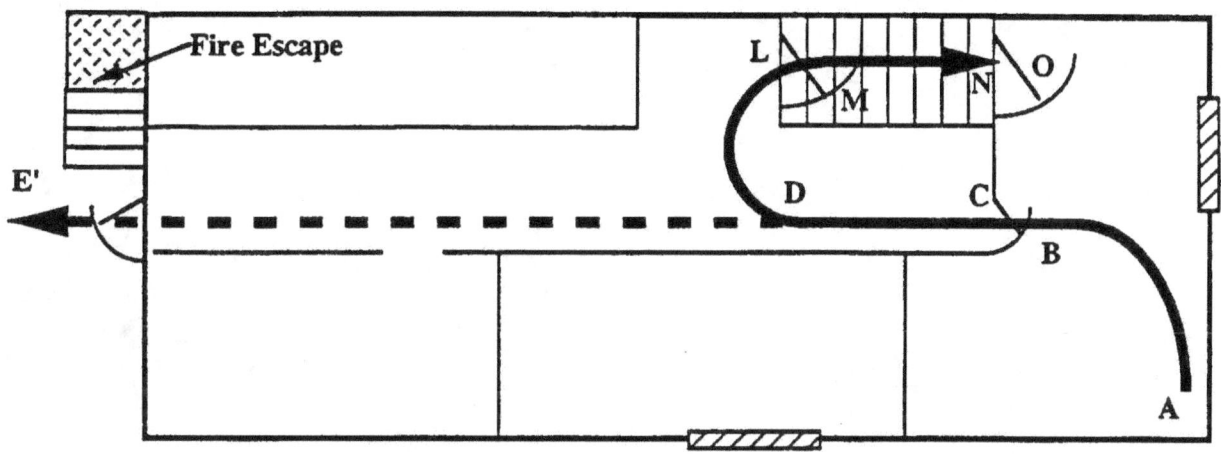

Plan View-Second Story

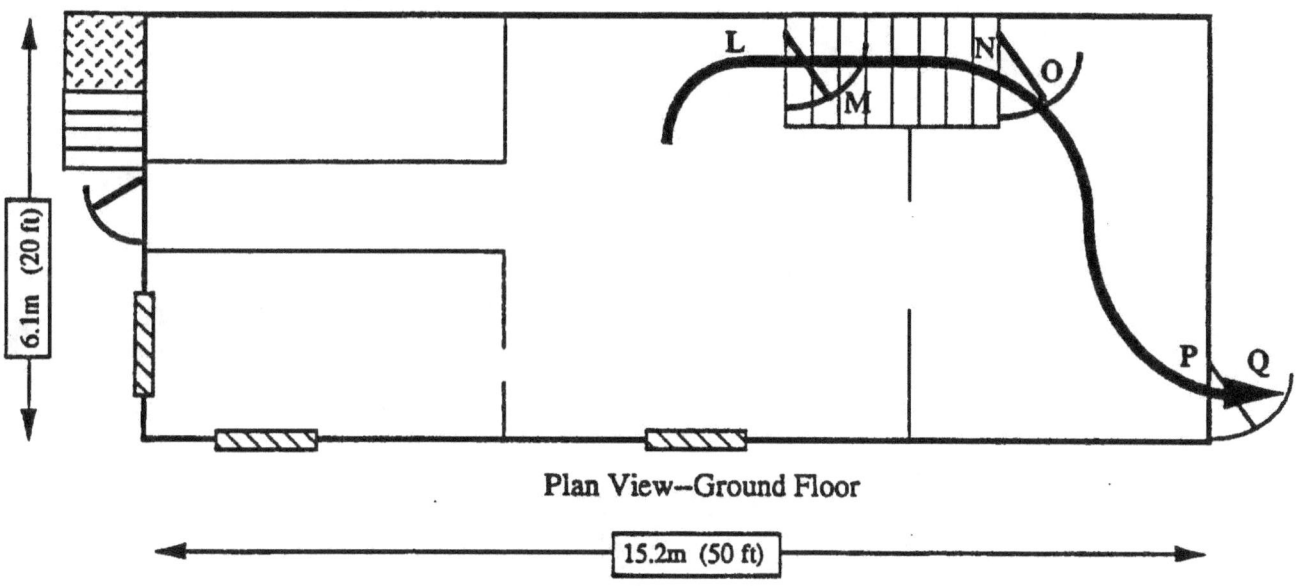

Plan View—Ground Floor

Figure D-2. Egress Paths from the Second Story of a Small Board and Care Home